LIVING
with
GOATS

EVERYTHING YOU NEED TO KNOW TO
RAISE YOUR OWN BACKYARD HERD

*

Margaret Hathaway

Photographs by **Karl Schatz**

*

LYONS PRESS
Guilford, Connecticut
An imprint of Rowman & Littlefield

W9-AZD-450

FOR BEATRICE, WHO HAS NOT KNOWN LIFE WITHOUT GOATS

Copyright © 2010 by Margaret Hathaway
First Lyons Press paperback edition 2012

ALL RIGHTS RESERVED. No part of this book may be reproduced or transmitted in any form by any means, electronic or mechanical, including photocopying and recording, or by any information storage and retrieval system, except as may be expressly permitted in writing from the publisher.

Lyons Press is an imprint of Rowman & Littlefield.

Photographs © 2010 Karl Schatz

Project Manager: Jessica Haberman
Text design: M. A. Dubé
Layout: Ann F. Courcy

ISBN 978-0-7627-8440-0

Printed in the United States of America

Distributed by NATIONAL BOOK NETWORK

The Library of Congress has previously cataloged an earlier edition (hardcover) as follows:

Hathaway, Margaret.
  Living with goats : everything you need to know to raise your own backyard herd / Margaret Hathaway ; photographs by Karl Schatz.
     p. cm.
  Includes index.
  ISBN 978-1-59921-492-4
  1. Goats. I. Title.
  SF383.H38 2009
  636.3'9–dc22

                                                                          2009007246

CONTENTS

ACKNOWLEDGMENTS

This book has been the work of many, but any mistakes are entirely our own. My husband and I are indebted to everyone who has helped shape our life with goats, but a few warrant special thanks: Our gratitude will forever go to our families, especially Jeanine Hathaway, Steve Hathaway, and Bruce and Nancy Schatz, for encouraging us to follow our agricultural dream. Without their support, our lives might never have included goats.

We wish to thank our editor Holly Rubino for her enthusiasm and her careful eye. Thanks also to Jessica Haberman and the whole team at The Lyons Press for putting together a beautiful book. Thanks to Jill Grinberg and her associates for shepherding us through another goat project.

On the home front, we wish to thank Jody Fein and Robb and Luisa Hetzler for their excellent goat- (and child) care. Thanks to Heather Rice-Sawyer and Jeannie Sawyer for all their help with the girls—this book wouldn't exist without them. Our goats are indebted to the team at Brackett Street Veterinary Clinic, especially to Dr. John Flood and Dr. Catherine Sanders for all things veterinary. Thanks also to John Ames and the folks at Ames Farm Center for never

Nine-year-old Erica Maser holds a one-week-old Alpine buckling kid at Ten Apple Farm in southern Maine. It's always good to have extra hands on deck during kidding season. Erica and her mom, Shari, came for a "farm vacation" from their home in Michigan to help out with the goat kids.

leading us astray. Phil and Robert Cassette have generously accommodated our schedule and answered myriad questions about breeding our goats. Ken and Janice Spaulding were kind enough to let us observe and photograph their Goat School. Dawud Uma shared both compassion and skill in demonstrating the proper technique for a *halal* slaughter. Leslie Oster's delicious goat creations continue to delight us. Our deep gratitude goes to Evin Evans, Caitlin Owen Hunter, and Wendy Pieh for their mentorship over the years, and especially for their extensive comments on the manuscript—you make us better goatkeepers. Thanks to José Azel and the great crew at Aurora Photos. A special thanks goes to everyone who welcomed us into their own lives with goats, both during our Year of the Goat and since—and allowed us to learn from and photograph them.

Finally, thank you to our human kids, Charlotte and Beatrice, for embracing goats with such gusto.

 A dairy goat climbs on an old car in Corinth, Vermont. Goats are enthusiastic climbers, always looking for high ground or a perch with a good view.

INTRODUCTION

Why Goats?

The morning begins with the gentle, rhythmic splash of milk streaming against the side of a stainless steel bucket. Through the barn windows, the sky lightens from deep, speckled cobalt to a pale, starless blue. The barn is suffused with the mingled smells of hay, sweet grain, and an earthy, animal pungency. Over the gurgle of ruminating bellies and the steady munch of oats, the bucklings in the kid pen bleat softly for their breakfast, and from his perch on the other side of the barn, the black rooster greets the day.

Perhaps this is the morning you imagine, the morning that encouraged you to pick up this book. Or maybe yours begins with the sun rising over a dewy emerald field and the herd eagerly approaching, knowing that you're bringing them grain and hay and fresh water. Or maybe it's you and a couple of pack goats out on the trail, yawning and stretching your limbs after snuggling together for warmth through a chilly desert night.

Life with goats can mean any of these scenarios. One of the great beauties of this incredible animal is its versatility. With proper care and attention (and even without), a goat can feed, clothe, and comfort you. It can thrive on one acre or one thousand and can adapt to virtually any climate. It can show you immense affection, nuzzling your hands and face, rubbing its flank against your hip and gazing up at you demurely. Other times, it can be the most stubborn, mischievous, and maddening animal in the barnyard—or, when you arrive home to find the electric fence down and the goats gone wild, the most frustrating animal in the cauliflower patch. Goats are gregarious herd animals, who love to be with companions. When their needs are met, they won't want to leave their haven, but they will want to explore it. One of the first lessons your goats will teach you: There is no fence strong enough to thwart their will.

Krusmynta, a Cashmere doe, is the queen of the herd at Wendy Pieh and Peter Goth's Springtide Farm in Bremen, Maine.

My own life with goats began in a New York apartment, where my husband and I dreamt of the country and fantasized about what animals would populate our barn. Five years and many miles later, Karl and I are raising our two daughters on Ten Apple Farm, a homestead in southern Maine where we keep a small herd of Alpine dairy goats, assorted poultry, and a large kitchen garden. After much research and exploration of the American goat industry, we chose dairy animals for a simple reason: We love cheese. But with any dairy herd you must cull, so we are by default also raising meat goats. And then there are the two wethers, hefty castrated males that Karl is training as pack animals. Another early goat lesson: They can become addictive.

We can no longer imagine our lives without goats, and I suspect that it will be the same for you. They nibble and nudge their way into every aspect of your life, from the time you set your alarm (earlier and earlier—there's always a hoof to trim), to when and where you vacation (the American Dairy Goat Association's annual convention? Fiber Frolic? Picking up that Boer buck in Texas?), to which novels you read in your spare time (try goatspotting in Barbara Kingsolver's *Prodigal Summer*, Charles Frazier's *Cold Mountain*, and Patrick O'Brian's seafaring sagas). Family stories will start to include goats: Our daughter Charlotte's first word was "goat," and when a week-old kid wouldn't eat, it was she, at age two, who coaxed him to the bottle. Family rooms are sometimes taken over; I can't think of a single goat person we know who hasn't, at some point, nursed a weak kid in their kitchen. In short, when you invite goats into your life, they make themselves at home. This is another reason to love them.

Living with Goats is the book I wish I'd had when I first became captivated by these animals. In the following pages you'll find practical information about goat husbandry, stories of life on the farm, and resources for answers to more detailed questions. This is by no means a comprehensive guide—for that I would direct you to the bibliography in Appendix C and the state extension offices in Appendix D in this book—but it sets out to answer the basic questions that a new farmer might have. Since Karl and I have only been at this a few years, these questions are fresh in our minds, and we hope that by gathering them here we can spare you some of our anxieties, frantic vet calls, and late-night Web searches.

While you're delving into this book, and certainly before you get any goats, my first serious piece of advice is to find a mentor in your community. No matter how many books you have on your shelf, nothing beats a reassuring voice, and sometimes, if possible, a set of more experienced hands. Goats are resilient animals, and when provided with good food, sturdy shelter, and hygienic conditions, healthy goats tend to stay healthy. But every so often you'll encounter a new situation that stretches your experience and you'll need sound advice, quickly and in the moment. Whether it's a trusted vet, a local rancher, or even the moderator of an Internet chat room on the other end of the line, you'll be glad to have an established relationship when the inevitable anxiety rears its head.

Life with goats can be a wild and wonderful ride. With a little preparation and forethought, you can avoid some of the common errors of new farmers (ourselves included) and give yourself the best experience possible. Have no doubt that keeping goats is a lot of work, but it is tremendously rewarding. Whether it's the glory of delivering your first kid, the dazzle of presenting friends with homemade cheese, or even the simple joy of a well-trimmed hoof, caring for your animals will give you a deep sense of satisfaction. Goats have changed my life in ways I never could have imagined. They've connected me to the land, to my food, and to my family. I wish you the same experience.

CHAPTER ONE
Things to Think About in Advance

When Karl and I first began to look into the possibility of life with goats, one of the animals' greatest attractions was their incredible versatility. The sheer multiplicity of their uses was inspiring. We were also, frankly, overwhelmed by the options. As livestock, goats can be raised for their milk, their meat, and the fiber of their coats. They can also be raised as pack animals, self-propelled weed trimmers, family pets, pedigreed show animals, and, at Ringling Brothers' winter quarters, as circus performers.

Though it sounds absurdly basic, the first thing you need to consider before buying goats is what you'd like to do with them. Are you looking for a dairy animal that will provide milk for your family? A pesticide-free brush clearer? A family project? Choosing a type and breed (or cross-breed) of goat early on will help ensure that the animal can meet your expectations. You wouldn't want to try milking a goat that has been bred to support a muscled, meaty frame, nor would you want to try shearing and spinning the hair of a dairy goat. Choosing the right breed will help you avoid such pitfalls.

Goats are very smart and, despite a serious stubborn streak, their intelligence makes them relatively easy to train. They are herd animals and prefer to have a companion, but with enough attention from you, they will consider your family to be their extended herd.

A three-month old Alpine buckling browses in the tall weeds of the goat paddock at Ten Apple Farm in Gray, Maine.

In our family, Karl is the lead goat in our barn and our herd will follow him anywhere. In the winter, he goes out on snowshoes and takes them for walks in our woods. He doesn't need to lure them with leashes or treats; they simply follow him wherever he goes. Because of their adaptability, goats will fit themselves into your life and will probably fill many roles simultaneously. Even so, before you bring them home, it's best to have a plan that takes a few key points into consideration.

CHOOSING A BREED

Goats were one of the earliest domesticated animals, and evidence of their life with humans can be traced back as far as ten thousand years. Though the animals probably originated in Central Asia and the Mediterranean, by the 1500s Spanish and Portuguese explorers were scattering goats across the globe. They were commonly kept on ships as a source of fresh milk and meat for the officers (there's documentation that goats were along on the *Mayflower*'s journey to America), and they

were often released onto islands in breeding pairs, thereby ensuring meat along the route for sailors—think Robinson Crusoe and his herd of wild goats.

Over the millennia, various breeds have been developed for different purposes. There are more than three hundred breeds worldwide—from the shaggy-coated, scimitar-horned Icelandic goat to the trim, compact Boer—but they can be divided roughly into three categories: dairy, meat, and fiber. This does not mean that a culled dairy kid won't make a delicious roast, or that an angora won't produce milk for its offspring. In fact, one of our does, Chansonetta, is an inadvertent Alpine/Boer cross, and though she doesn't give much milk, hers is by far the richest in the herd. Rather, each type of goat has been bred to excel in a certain area, and in choosing a breed, you'll be choosing a focus for your herd.

DAIRY GOATS

For the homesteader or small-scale farmer, dairy goats are often an ideal choice. Full disclosure: We raise Alpine dairy goats, so I'm a little biased, but their advantages are many. A good dairy producer from any breed can give you more than a gallon of milk a day, which is plenty for most families. The majority of dairy goats have been bottle-fed and raised in close proximity to humans, so in general they are friendly and personable, which makes them a nice addition to the barn. They are also generally disbudded, meaning that their horns are removed soon after birth, which makes them easy for the whole family to handle (and eliminates some parental anxiety if your toddler and your goats are both on the loose in the barn). If you're interested in getting children involved in 4-H, or in showing animals yourself, the American Dairy Goat Association's network of sanctioned dairy goat shows is vast, and there are dairy goat associations in almost every state, which will give you an instant community of more seasoned farmers to turn to for mentorship and advice.

A pair of Oberhasli dairy goats are contained behind four-foot woven-wire fencing at Elizabeth Kennerly's Devonshire Dairy Goats in Archer, Florida.

The drawbacks to dairy goats are few, but they are substantial. The first is the time commitment. When your goats are lactating, you or someone on your farm must be available to milk them twice a day. (We have heard of breeders who raise their goats exclusively for the show ring and prior to a show only milk them once a day. Given how large our goats' teats become by milking time, I can only imagine how uncomfortable this makes their poor goats!) At our farm, we split daily chores so that Karl does the morning milking and I do the evening, but even with a division of labor, milking the goats means that at least one of us has to be in the barn and ready to milk at 5:30 a.m. and 5:30 p.m. (You can start later in the morning—conceivably you could milk at noon and midnight—but the important thing is consistency.) Milking definitely ties you to your home and your barn, which can feel wonderful but also a little confining. You can't stay late at work, or decide to grab an early dinner and a movie, and unless you have a trusted farm-sitter who knows your routine, while your goats are lactating—usually in the spring, summer, and fall—it's virtually impossible to take a vacation. If you have difficulty following a routine, or have unpredictable demands on your time, dairy goats might not be the right choice for you.

Another potential drawback, which can be looked at in a different light as an advantage, is the need to breed your goats to keep them in milk. Like all mammals, goats need to give birth in order to begin lactating. (It seems almost insultingly fundamental to mention that, but you'd be surprised by how many people have asked us if we're milking our males.) After months of milking, the volume of milk begins to decrease; goats are "dried off" (allowed to stop lactating) and bred again to ensure good milk production. Many commercial herds encourage extended lactation, in which the does are milked for two years between breedings, and at our farm, we've experimented with breeding our does in alternate years so that we're never without milk. Inevitably, however, the does will need to be bred again. In practical terms, this means that when you're ready to breed them, generally in the fall, you will watch your goats for signs of ovulation and then do one of three things: put a buck in with your does, take them to another farm to be bred, or artificially inseminate them. Five months later, if all goes well, your pregnant does will each give birth to between one and four kids, which you can then decide to raise, sell, or slaughter.

An Alpine dairy goat gazes at the camera at Rick and Lora Lea Misterly's Quillisascut Farm in Rice, Washington. Rick's favorite thing about goats is how affectionate they are. "When you see them in the morning, they begin your day in the most amazing way and make you happy."

I'll go into more detail about breeding and kidding later in the book, but it's an essential part of raising dairy goats, even on the smallest scale, and it should be considered at the outset. There's something magical about helping to deliver another creature, and it's an incredible experience to share with your children. But kidding season can also be fraught with anxiety and can offer unexpected lessons in mortality. In the majority of cases, goats give birth easily and need very little assistance from you, but breeding is still not something to be undertaken lightly. If you can't be available around the clock while your does are kidding, dairy goats may not be the right choice for you.

If you do decide on dairy goats, your next choice is which breed to raise. There are currently eight breeds recognized by the American Dairy Goat Association, and if you're interested in showing your animals in competition, they must be registered as one of these breeds. If you're not interested in raising breeding stock, or want to experiment with cross-breeding, you can have a wonderful experience with goats. But you'll still want to choose animals from these breeds, which have been bred for generations specifically to produce a large volume of rich milk. I'll list them in alphabetical order to avoid any suspicion of bias (though you might notice that Alpines come first . . .).

Alpine

> These goats have erect ears and come in many different color patterns. They originated in the Alps and were brought to the United States in 1902. Alpines are very popular with commercial goat dairies because they are considered high-volume milkers. According to the USDA, Alpines have the highest annual average milk production of any of the dairy breeds.

La Mancha

> La Manchas are easily recognized because they have exceptionally small ears. The only breed to originate in the United States, they were developed in the 1930s as a cross between a handful of mysterious short-eared does and some registered Swiss and Nubian bucks. Like Alpines, La Manchas come in a variety of color patterns. They are generally considered to be very friendly and gentle, with pleasant, even-tempered personalities.

Nigerian Dwarf

Nigerian Dwarf goats are originally from West Africa and are recognizable by their small stature. Like Alpines, they have erect ears and their coats come in many different colors and markings, but their average height is just under two feet. First introduced at zoos in the United States in the 1980s, Nigerian Dwarf goats don't give as much milk as their larger dairy counterparts, but their milk is exceptionally rich in butterfat, making them an unorthodox but valuable addition to commercial herds. Specially scaled milking equipment to use with this breed is available from goat supply companies. Nigerian Dwarf goats are also distinctive in that they are fertile year-round, and can be bred in any season. The most recently recognized breed in the United States, the Nigerian Dwarf registry has been accepted since 2005 by the American Dairy Goat Association.

Nubian

Nubians may be the most popular goat breed in the United States, and are recognizable because of their distinctive long ears and prominent convex noses. Originating in Africa, Nubians have been in America since 1909. They have large frames and have a reputation for being stubborn and very vocal, and for having strong personalities. Nubian milk has a very high butterfat and protein content, which makes it excellent for cheese making.

Oberhasli

Oberhasli goats are recognized by their distinctive "chamoisee" markings: dark reddish brown or "bay" coats, marked by black stripes on their face and underbelly. Originally a Swiss breed, Oberhaslis were for many years considered Alpines by the American Dairy Goat Association and were not recognized as a separate breed until the 1960s. There are relatively few registered Oberhaslis in the United States, and even fewer in commercial herds.

Oberhasli dairy goats lounge around their pen at the 2004 American Dairy Goat Association's National Show. The goats belong to Dave and Peg Daubert of Watertown, Minnesota, who have had many wins at the show with their Tonka-Tails Oberhaslis.

Saanen

Saanens are another Swiss breed, originating in the Saane Valley of Switzerland and introduced to the United States in 1904. They are recognizable by their pure white coat, erect ears, and concave nose, and by the delicate freckling of their udders that occurs when their fair skin is exposed to the sun. Saanens are very popular with commercial breeders because of their large size, calm temperament, and high average milk production.

Sable

Sable goats are simply Saanens that are not white. While they share the same traits as Saanens in every other respect, their coloring is the result of two recessive genes, and they can consequently occur in any Saanen herd. The first American Sables arrived with Saanens in the early 1900s, but until the last few decades were considered deviations from the breed. Registered Sables can be any color except white, and have been recognized as a distinctive breed by the American Dairy Goat Association since 2005.

Toggenburg

Sometimes referred to as "Toggs," the Toggenburg is a Swiss breed recognizable by the distinctive white "reverse badger" facial markings on its typically brown or light brown coat, as well as its white lower legs and tail. Introduced to the United States in the early 1900s, the Toggenburg is the oldest registered breed of goat in America.

MEAT GOATS

Unlike spring lambs, which are often raised over the summer to fill the winter freezer, very few Americans raise meat goats purely for their family's consumption. While domestic consumption has never been higher, Americans are still learning to love—and to cook—goat meat. If you're interested in buying kids in the spring, keeping them for brush removal over the summer and fall, and using them for hearty stews in the winter, you'll enjoy a low-impact livestock that makes relatively few demands on your time and rewards you with delicious, healthy meat. My guess, however, is that if you're considering raising goats for meat, you have a slightly larger scale in mind.

A word of caution at the outset: As with dairy animals, there are registries for various breeds of meat goats. When Boer goats were first introduced to the United States in the early 1990s, registered breeding stock fetched tens of thousands of dollars per animal. Since then, the number of farmers raising meat goats has steadily risen, as has the demand for goat meat. The price of breeding stock, however, has fallen. If you are interested in breeding goats for the satisfaction of watching your animals meet breed standards in the show ring, you will enjoy raising registered Boers. If, however, you are interested in breeding goats to turn a quick profit, you will be disappointed.

There are a variety of factors to consider before deciding to raise meat goats. The first is scale: How much land do you have and how many goats would you like to put on that land? The general recommendation from seasoned farmers and state extension officers is that six to ten goats can be kept per acre year-round. Using that calculation, just a few acres will give you space for a pretty substantial herd, though the more concentrated the goats are on the land, the more management they will require. More on that later.

A second consideration is whether you're planning to raise animals predominantly for slaughter, or if you're concentrating on breeding stock. If you've identified a market for goat meat in your area—a large population of African, Latin American, or Central Asian immigrants, perhaps, an established farmers' market, or simply the growing number of consumers demanding meat from naturally raised and humanely slaughtered animals—and you've found a small slaughterhouse to work with, then it might make sense for you to raise unregistered animals for slaughter. If this is the case, you may choose to begin with a foundation herd of well chosen animals—does and possibly some bucks—with the intention of breeding them yearly and selling their offspring. You may decide to keep female kids until you've increased the herd to the desired size, or to only keep replacement kids when does are "retired." Either way, you'll need to remember that your herd will double at least briefly each kidding season, and keep that number in mind when you're calculating how many goats your property can support. You'll also want to make sure that you have a targeted marketing plan and that you have

the ability to comply with the United States Department of Agriculture's regulations for slaughtering and butchering meat.

 Mother and daughter Boers, Sophie (right) and Annie, soak up the autumn sun at Ken and Janice Spaulding's Stony Knolls Farm in Saint Albans, Maine. The Spauldings purchased a pregnant Sophie from a breeder in Texas. According to the Spauldings, Sophie is a fabulous mom, and Annie has turned into a fine doeling from the excellent care.

If you're interested in showing your animals or raising breeding stock, then you might want to start very small, with a handful of registered goats that conform to breed standards. In the United States, there are currently registries for Boer, Kiko, Myotonic, and Savanna goats, with Boers and Kikos making up the greatest percentage of registered meat goats.

Boer

Originating in South Africa, Boers are sometimes referred to as the "Black Angus" of meat goats. Compact and muscled, Boers have a stocky, trim frame, convex nose, long ears, short horns, and distinctive chestnut and white coloring. Introduced to the United States in 1991, Boers were initially cross bred with Nubians to create "percentage" animals (goats of mixed breeds with documented lineage) and to increase breeding stock. After several generations of well-documented breeding to other Boers, the offspring of these percentage animals conformed physically to breed standards and the goats were officially considered to be Boers. There are currently three Boer registries in the United States, and each association provides resources for its members, sponsors sanctioned shows, and conducts youth programs. Before joining a Boer goat association, it's a good idea to see which have active chapters in your region.

Kiko

Kiko goats originated in New Zealand, where domesticated bucks were bred to feral does to create a new breed of large, hardy, parasite-resistant goats. Massive, with shaggy white coats and wide scimitar horns, Kikos were introduced in the United States in the early 1990s. There are two Kiko registries in the United States, and their associations provide members with resources for breeding and business management. There are currently no specific shows for Kiko goats, though they are sometimes shown in a broad "meat class" at fairs.

In addition to registered breeds, Spanish goats are another popular meat breed. Taking their name from the Spanish explorers who introduced them to the New World, Spanish goats are often called "scrub" or "brush" goats and are found throughout the country, though they are especially concentrated in the South and Southwest. Spanish goats are generally large-framed and hardy. The Spanish Goat Association is dedicated to the preservation of the gene pool of original Spanish goats. The organization does not hold shows or look for specific physical characteristics, and its resources for members are focused on education and breed conservation.

Once you've chosen a scale and a breed, the needs of your meat goats will change with the seasons. While they're breeding, generally in the fall, and kidding, generally in the spring, they will demand a lot of attention. As with dairy goats, you will need to make decisions about breeding and be available around the clock when your goats kid. Unlike dairy goats, with meat breeds you will generally not need to bottle-feed the kids, and you won't remove their horns, since it's customary to leave the horns of meat goats intact.

The rest of the year, your herd of meat goats will need management—hoof maintenance, feeding, grooming if you plan to show them—but you have quite a bit of latitude when it comes to demands on your time. Some breeders keep small, intensively managed herds, with extensive documentation of the goats' feeding schedules and weight gain. Others keep less detailed records and take a more laissez-faire attitude. As long as your animals are healthy and provided with ample food and appropriate shelter year-round, one of the wonderful aspects of choosing to raise meat goats is that they allow you to create your own path.

FIBER GOATS

If you spin, knit or weave, fiber goats might be the right animals for you. There are two breeds of goats recognized for their exceptional fleece: Angora, from which mohair is shorn, and Cashmere goats, which develop a fine undercoat of cashmere that is then combed out. Both breeds have modest frames and impressive horns, and animals culled from fiber herds are commonly sold for meat. Before the abolishment of the American government's mohair subsidy in 1993, South Texas was home to a thriving Angora goat population—many of the goats remain, as well as some wonderful *cabrito* barbecue joints.

Unlike dairy or meat goats, Angoras and Cashmere goats do not have to be bred regularly, but the quality of the fiber they produce will degrade over time and is best in their first few years of life. Before choosing to raise goats for their coats, it's a good idea to investigate whether there are other fiber goat breeders in your area, and to talk to them about the regional opportunities for the sale of animals (live and for slaughter), as well as the possibility for cottage-industry scaled processing. Shearing mohair, combing cashmere, and cleaning and spinning any fiber can be a wonderful Zen-like project, but it is time consuming. When thinking of the size of your herd, make sure to anticipate the time it will take to collect and process the fiber. Though they'll thrive in many climates, it's always a good idea to take your environment into consideration: If you live in a hot, humid region, raising goats for fiber may present special challenges.

Some notes about fiber breeds:

Angora

Angora goats are generally white, though some breeders are now deliberately raising them in silver, black, and deep, mocha browns. Their stature is small (in fact, Angoras have been bred with Pygmy goats to produce "Pygoras"), their horns are long and arced, and their hair is thick and curly. Originally from Turkey and the Anatolian peninsula, the name angora is actually a corruption of the word *Ankara*. Their mohair, which is softest when they're young, is shorn twice yearly, when a lock can be stretched to four to six inches.

Cashmere

Though almost any goat can produce some amount of cashmere—the soft undercoat that is combed or shorn from below an animal's coarse, outer coat—Cashmere goats have been specifically bred over time to produce a greater than average amount of fleece. Originating in Central Asia—the name comes from the word *Kashmir*—cashmere goats are raised throughout Asia. China is the world's largest producer of cashmere. Cashmere goats naturally produce their fine undercoats between the summer and winter solstices, and shed it in the spring, generally producing about four ounces per goat per year. Since any goat can produce cashmere, for many years there was no breed standard for a "purebred" Cashmere goat. Recently, however, the North American Cashmere goat was recognized: a mid-size animal with a long, shaggy coat, and broad, scimitar horns. To meet breed standards, its fleece must have a diameter of nineteen microns, and be at least one and a quarter inches long. Cashmere goats come in all colors, from white to silvery grey to black.

OTHER POSSIBILITIES

If you're interested in keeping goats as pack or cart animals, weed eaters, or simply pets, you can raise virtually any breed. Dairy wethers—neutered males—are often used as pack goats because of their large frames and gentle temperaments, but if it's well trained and can carry your load, there's nothing to prohibit the use of either another breed or a doe. John Mionczynski, widely considered to be the "Father of American goat packing" because of his expertise and authorship of the goat packing classic *The Pack Goat*, tells stories of extended hikes where he's eaten solely what he's foraged, supplemented by the milk of the doe that carried his gear.

Because they are browsers, all goats are exceptional brush clearers. Lani Malmberg, who rents out her herd of 1,200 Cashmere goats for land maintenance throughout the West, often lectures in a dress made from cashmere she combed from her herd! Pygmy goats, a breed mentioned earlier in passing, are small goats—full-size adults are generally not taller than two feet—that have been bred to be pets, though they can produce modest amounts of milk. Many other breeds which we haven't discussed—Golden Guernsey, Arapawa Island, and Myotonic (also called "fainting") to name a few—are making their way into herds in the United States. There are as many ways of living with goats as there are people who live with them, and once you've decided what you want to do with your goats, the possibilities are endless.

ACQUIRING YOUR GOATS

Now that you've chosen a type and breed, the next step is getting some goats. Even under the best circumstances, it can be daunting to choose animals, and it's especially hard if you're not sure what to look for. No matter what the goat's age, gender, or registration status, if it's at all possible, I highly recommend bringing a more seasoned farmer along on your first forays into goat buying. A neutral party who can act as a guide will help you avoid feeling at the mercy of breeders, and having a second pair of eyes evaluating the animals will give you a clearer view of the goats. It's easy to fall in love with a kid's markings or bouncy temperament, but that quick

affection can sometimes keep you from seeing an animal's imperfections.

 A dairy goat explores a piece of wood with his tongue during a goat packing demonstration at the American Dairy Goat Association's 2004 National Convention in Albuquerque, New Mexico. Goats are naturally curious, often exploring objects and surfaces by nibbling or chewing.

When Karl and I chose our first kids, we fell in love with a chocolate-colored doeling who was born while we were at the farm. I toweled her off and bottle-fed her, and we were so smitten that we didn't notice her stocky build, snub nose, and airplane ears. Caitlin Hunter, the friend from whom we bought our goats, hadn't planned to keep many of her kids so she bred some of her Alpine (dairy) does to a Boer (meat) buck to bulk them up for auction. By the time we all figured it out, Karl and I were too attached to the kid to see her go to auction and now Chansonetta—a little tank who eats a lot but gives us very little (but very rich) milk—is queen of the barn. I like to think that experience has made us wiser.

It's optimal to buy animals directly from the farm, and even better if you know the farmer. A quick look in the barnyard will give even the most inexperienced goatherd an idea of the condition of the animals. Look for lustrous coats, a spry gait, and good body condition—not too fat or too thin. They shouldn't have a significant over- or underbite—often referred to as parrot mouth and monkey mouth—either of which can make feeding tricky. Don't be alarmed if they don't have front teeth, however. Goats have teeth in their lower jaw, but an upper dental palate (actually, you should be alarmed if they *do* have front teeth!). You'll want to reconsider if the goats have runny noses and eyes, evidence of scours (chronic diarrhea), scabby spots, brittle or patchy coats, or obvious trouble walking. These are all signs of illness and disease. As much as you may want to remove animals from an unfortunate situation, unless your aim is to rescue livestock—which is a perfectly wonderful thing to do—you won't want to add sick or injured animals to your herd. In most cases, if you know the farmer and are willing to buy his goats, odds are you agree with his philosophies about raising animals and you'll be able to consult with him if there are any problems down the road.

If you're buying registered (purebred) animals from a distance, make sure that you see their documentation before you go to great lengths arranging transportation. Likewise if you're buying a doe that's already been bred, make sure that you not only have information about the sire, but also precise dates of breeding so that you can plan accordingly.

If you're buying an existing herd in a dispersal sale from someone who is getting out of the business of raising goats, get as much information about the herd as possible. Ask questions about the dynamics within the group—who is the alpha goat? is there an order in which they eat? will they come to you for grain or treats?—as well as documentation of vaccinations and any medical treatment and contact information for prior vets. It can also be helpful to know why their previous owner is selling them.

You may initially feel awkward asking questions about the health of someone's herd, but don't. Serious breeders and farmers will never take offense. Karl and I were shy at first to ask about vaccinations and the incidence of illness when we were looking at goats for both purchase and breeding. However, Phil Cassette, a farmer in Maine whose family has been in dairy goats for decades and to whose bucks we've bred our does, assured us that it's not impolite. In fact, he said, our questions let him know that we're responsible about the health of our own herd.

In all of this, I'm assuming that you're buying goats from a farm or a breeder. As a new farmer, I would emphatically discourage you from buying goats at livestock auctions. The majority of goats being sold at livestock auctions have been culled from their herds and are being sold for meat. It is certainly possible to find a healthy, desirable goat at an auction—our beloved Chansonetta and both of our wethers could easily have ended up in the sale ring—but the format of the sale makes it difficult, if not impossible, to ask questions of the animal's previous owner. Why tempt fate? The more information you have, the easier it will be to manage your animals and the greater the chance that your life with goats will be enjoyable.

Getting Them Home

You'll need to think about transportation for your goats before you actually choose them. Your needs will vary depending upon the size of the animals and the number you're bringing home, but in all cases you'll want to move them in something that's enclosed. While a pickup with a deep bed will work for some small ruminants, goats will immediately hop out. Ideally, you'll have a livestock trailer or a pickup with a closed top on the back available to you when you need to move your goats. If you're considering showing goats, access to a trailer is essential.

With a small herd, we've managed to make due without a trailer so far, but there are times (when the back of my Subaru smells like goat urine, for instance) when I desperately want one. Our favorite makeshift options include a large dog crate in the back of the car for transporting babies and small breeds. We brought our four kids home when they were about a month old, and they huddled together in a hay-lined plastic dog crate for the drive home. If you have an SUV and a pressure-mounted pet barrier (usually for dogs, these are sold at most pet stores), you can put down a piece of plywood or a tarp in the back, cover it with hay, and you should be able to transport an adult female for breeding or vet visits. If you give them a little snack for the road, most goats will settle down and relax after a few minutes. Bucks, wethers, and large does, however, need equipment that's more substantial, especially if they're going to be traveling for more than a couple of hours.

CHAPTER TWO

Where Your Goats Will Live

Goats, like people, prefer a dry, cozy home. There are many ways to give them this, and I've seen goats thrive in shelters that span the range from a three-sided shed to a snug new barn with radiant heating in the floors of the kidding pens. The most important thing to the goats is that their shelter is free from dampness and drafts. Beyond that, the choices are up to you.

One of the first things you'll want to consider, when thinking about a home for your goats, is your own convenience. If you're bringing the animals into an existing barn, some decisions—like where the barn is in relation to your house—have already been made. If you're building a shelter and deciding where to place it on your property, you'll want to think a lot about proximity. With dairy goats, keep in mind that you'll need access to the animals at least twice a day, and as the seasons change, that first morning milking may be before daybreak. Depending upon the climate, you don't want your barn so far away that you wake up dreading a dark trudge through the snow.

On the other hand, you may not want a barn full of animals—with their droppings and their noise—directly in your backyard. Especially if you have enticing gardens, you

Three-month-old Alpine doeling, Flyrod, stands on top of a dog house shelter. The authors found the dog house for free on the side of the road and nailed some boards to the top so the goats could climb the roof. At three months, all four of the herd's kids could fit inside, although the goats quickly outgrew the structure.

might want some distance between your home and your barn; I've never known a goat that didn't stop to eat the flowers, even when it was

 Dairy goats peer out of a barn in Corinth, Vermont.

on the lam and its owner was in hot pursuit. If you live in a warm climate, goat bedding, no matter how frequently it's changed, can become pretty pungent.

The most important thing is to think through where you want to place your goats and to put them on a part of your property that makes sense to you. And keep in mind that no matter what you're doing with them, if there's any kind of emergency or great goat escape, you'll want to be able to get to your animals quickly.

When Karl and I bought our home, a sprawling farmhouse with an attached, two-story barn, one thing we loved about the house was the direct access to the barn from both floors. During the long Maine winters, when the days are short and the snow is deep, we don't even have to go outside to do our morning chores.

And year-round, if we hear a noise in the middle of the night, we can just go past the bathroom and through the door at the end of the upstairs hall to peek down and check on the critters. When the hayloft is full, the whole house smells like sweet clover. The drawback to the attached barn is the inescapable cacophony of chickens and bleating of goats, and the constant, fragrant reminder when we need to clean out the pens. But in our climate, these seem a small cost for the convenience in winter.

Once you've decided where to put your goat shelter, its requirements are pretty basic. Other than keeping them free from moisture and drafts, a goat shelter should provide the animals with shade during the heat of the day, protection from rain and snow, and, if at all possible, soft, comfortable bedding. Many farmers—especially those with dairy animals—choose to feed their goats inside, often during milking. If this includes you, you'll want stationary hay feeders and either a shallow trough or a space for grain pans. If you have dairy goats, you'll also want space for a stanchion (two bars that pivot and lock over a goat's neck to keep it in place) or milk stand. If you're planning to breed your goats, you'll need at least two pens so that you can separate does when they're kidding. Some farmers build sleeping plat-forms—raised "beds" about two feet off the ground—for their goats, while others simply provide clean, regularly changed shavings or straw. Beneath the shavings, a scattering of powdered agricultural lime will help dry out the barn floor and control flies in the barn by disrupting their breeding.

There are as many variations of structures that meet these needs as there are goat farmers, and I would encourage you to experiment. Ideally, your goat's home will open out onto their field or paddock, but if this isn't possible, there are ways to make any situation work. It's not necessary for all of your animals' needs to be met under one roof. There are many farmers, our family included, who have a few structures to satisfy their goats' needs: In our case it's a barn in which they sleep, eat, and are milked, and a more rudimentary three-sided shelter in the field for shade and protection from the elements during the day. Other farmers we know have a separate milking parlor or distinct pens for eating and resting.

When designing your structure, you'll need to think carefully about where to feed your goats. Contrary to myth, goats can be pretty finicky, especially about their food. A system that may seem ideal to you in theory might be completely rejected by your goats for reasons that seem entirely whimsical. While the rest of our herd makes a beeline to their evening grain rations, no matter where we place them, we have one goat that won't eat from a pan on the floor. If we make sure that at least one pan is resting on something higher up—even a humble cinder block—she's as hungry as the rest. But if we put all the pans on the floor, she'll skip the grain course entirely and head straight for the hay feeder. You can come up with a plan that is wonderful in the abstract, but don't be discouraged if once you introduce your goats, they make their own suggestions.

We started out with our goats in a converted horse stall and in just the few years that we've had them, their shelter has gone through at least half a dozen goat-instigated "renovations." They've broken their Plexiglas windows, worn down posts with their horns, snapped the gate latch more times than we can count, and even figured out how to sleep in their hay trough. We once came home to find the goats romping around the barn in a flurry of shredded papers. These were old bank

records that we'd been storing in a plastic filing bin, now in pieces on the barn floor. Some of our best ideas have proven no match for their hooves and horns. While it can be frustrating in the moment, it's part of what makes raising goats a constant adventure, and figuring out how to keep everyone happy is part of the fun.

BUILDING A SHELTER

Once you've decided where to put the goats, it's time for construction! Whether you're building a new shelter or repurposing an existing one, there are wonderful books and Web sites out there with more detailed instructions and plans than I can include here. A list of our favorites is compiled in the sidebar on page 24.

No matter how elaborate or basic your shelter, make sure that it is as sturdy as possible. Goats are notorious for their skills at dismantling: They rub, they butt, they kick, they climb, and they jump. In our experience, if you're considering a light canopy or tent for easy, portable shade, your goats will likely have it upside-down and trampled within the hour. Likewise, any kind of flimsy construction is a gamble: During a spell of exceptionally hot and clear summer days, we made a lean-to out of two-by-fours and plywood for extra shade. Our goats used it as a ramp, running up the side and jumping off the top until the plywood cracked and would no longer support them. After much trial and error, we now use thick boards, many posts, and heavy screws in all of our goat constructions.

We also use two layers of Plexiglas in all windows at goat-level. They've still managed to break it, but it doesn't shatter and the pieces are large enough that we can easily pull them out of the bedding. We've heard of other farmers covering glass windows with a protective layer of heavy wire mesh. If you're designing your own shelter and it's at all possible, placing windows above the height of your goats—at their tallest, while stretching up and standing on their back hooves—is the safest option. Goats do not like the dark and it's best to give them natural light if you can, but it's also essential to keep their safety in mind.

Cashmere goats are tied up at feeding time in an open ended barn at Wendy Pieh and Peter Goth's Springtide Farm in Bremen, Maine. Sassy Girl, in the foreground, is a major doe in the herd and has produced several of their best offspring.

RESOURCES FOR BUILDING ANIMAL SHELTERS

How to Build Animal Housing, by Carol Ekarius, is a wonderful survey of basic structures for the barnyard. We've used her plans for all kinds of things, including our goats' milk stand and our chickens' roosts and nesting boxes. Simple instructions make everything seem possible and encourage personal embellishment. Our copy is flagged with dozens of sticky notes.

Building Small Barns, Sheds & Shelters, by Monte Burch, is another classic resource for rudimentary building instructions and plans for farm structures. Designed for the small farmer and homesteader, it not only includes plans for animal housing, but also for things like a smokehouse and root cellar.

North Dakota State University's Extension Office maintains a Web site (ag.ndsu.nodak.edu/abeng/plans/index.htm) that includes comprehensive plans for all manner of agricultural buildings. Plans are printable and free, and though there's nothing specific to goats, there are many projects that can be adapted for their needs.

WINTER SHELTERS

 Flyrod wades through powdery snow outside the barn at Ten Apple Farm in southern Maine. In the winter, goats can be an entertaining addition to hikes and snowshoe trips, sometimes even in high snow.

If, like us, you make your home in a part of the country where winters are long and harsh, an appropriate cold-weather shelter will be a necessity. Goats are hardy and can withstand extreme conditions, but they don't like to. Given their choice, your goats will probably spend a lot of the winter indoors, so it's important to give them, in addition to snug quarters, a little entertainment. Goat packing guru John Mionczynski plays the accordion for his goats, and I've been known to sing and dance for the goats during evening chores (to the embarrassment of my family, our wether Joshua loves my appalling renditions of classic Jackson 5 tunes), but when I say entertainment, what I mean here is simply something on which to climb. An elevated platform, stacked cinder blocks, or even an old tire will give the goats a little something to keep them occupied.

 A Pygmy kid peeks out from under a tractor tire shelter at River Valley Farm Stand in Oxford, New Hampshire. Recycled objects such as tires, large wooden spools, and cement blocks can provide shelter and entertainment for goats.

Because our barn is on a stone foundation and the goat pens are somewhat elevated, we do two things to prepare for winter: We let the bedding build up a little in the fall to create a layer of insulation, and we drag in a pyramid of cinder blocks for the goats to climb. It's not much, and we supplement it with outdoor hikes as much as we can, but when the snow is up to the goats' withers (the ridge between the shoulder bones) for three months of the year, we have to give them something to scale.

Winter is the only time of year that we aren't meticulous about changing the goat bedding. Goats are fastidious, and they seem to resent it when their pens become untidy. Aside from basic issues of sanitation, your goats will be happiest in neat quarters. We use pitchforks and shovels because our herd is small, but many farmers, especially those with substantial herds, use tractors or Bobcats. If you plan on the latter—which will spare your back and certainly shorten the amount of time spent on chores—make sure that your pens are tall and wide enough to maneuver the machine. Also, if you're designing your pens, make waste removal as easy as possible; a straight line between the pen's entrance and an exit, trap door, or removable panel works best.

By winter's end, shavings, manure, and stray bits of hay have felted themselves into a thick mat. We find that separating the layers with a pitchfork and then shoveling them out the door works

Sable Saanen and La Mancha kids play on the galvanized roof of a shelter at Vern Charles's Charles Ranch in Delta, Colorado. Simple three-sided structures like this one can provide shelter and a place to climb.

best. Without first peeling the stuff apart, it is impenetrable. When you do summon your courage for the early spring barn cleaning, make sure to choose a day with cooperative weather: You'll want as much ventilation as possible, and rain or heavy winds will make your chores that much harder.

Goat manure makes excellent fertilizer, and before we clean the barn each spring, we till some of last year's pile of composted bedding into our garden. By that point, the sharp whiff of ammonia is gone and it's simply a rich, steaming pile. We do the same in the fall, when we till the garden under, using the first frost as a reminder to do a serious barn cleaning before winter comes. Stray seed heads from undecomposed bits of hay have given us some interesting weeds—including a beautiful pale pink poppy whose seeds we collected and saved—but our soil gets richer every year. Tying your barn cleaning to the garden schedule has the added advantage of discouraging procrastination; you can't put off changing the goat bedding if it will disrupt your first planting or your final harvest.

CHAPTER THREE

The Eternal Problem of Goat Fencing

If goats had an articulated mantra, it might be the words of Cole Porter, "Don't fence me in." Goats are smart and curious, and they seem to understand that in a pinch they can be pretty self-sufficient. This makes them probably the most difficult domesticated animal to contain. Especially if you've brought home an adult goat with a lifetime of bad habits, maintaining adequate fencing may be the most challenging aspect of your life as a goatherd. Just as goats will batter and pry at their shelter, they will heap on as much abuse as their fences can withstand. Even the most docile animal will occasionally rear up on its hind legs to stretch against the fence, rub the posts with its head and withers, or, if startled, to attempt an escape with a running leap. If there's a power struggle within the herd, frequent hip-checks and jostling often make the fence an early casualty. And if there's something interesting on the other side—an apple sapling, perhaps, or something tall, like your car, to climb on—nothing will stop a curious goat from wiggling through.

Our property was entirely unfenced when we bought it, and Karl and I looked at a lot of fencing options before we brought our goats home. We found advocates of all manner of goat containment, from tethering—attaching goats to a post in the ground with a chain—to tall, solid wood fences. Tethering, it should be noted, is an unsuitable method of containment for goats. Not only are the animals unable to pursue their natural inclination to wander and browse, but they are also

Tippy, a mixed dairy breed wether, peers over a woven-wire fence at Frank and Carol Webber's Highland Badger Farm in Ellison Bay, Wisconsin. Goats are notoriously hard to keep fenced in. Woven-wire fences, usually reinforced with strands of electrified wire, are often the fences of choice.

at risk for tangling and choking, and are especially vulnerable to predator attack.

What made the most sense to us at the time, and what we continue to use, is portable electrified fencing. Nothing will keep all goats in, all of the time—goodness knows, we've had our share of escapes—but we find that the steady click of the fence discourages regular breakouts while offering reasonable protection against the predators in our area.

The trade-offs are mainly the repeated expense of replacement batteries and the extra management required as our children get older and go exploring on our land. For the former, we contrast the financial outlay with the cost of property and losing neighborly goodwill if the goats were to wander regularly. For the latter, we've established family rules about turning off the fence whenever we're outside.

A Toggenburg dairy goat sneaks under a gate at Pleasant Valley Acres in Cumberland, Maine. For goats, the grass is always greener on the other side of the fence.

Establishing Good Habits

I'll address specifics about fencing a bit later in this chapter, but first I want to talk a little about training your goats. If you're acquiring older animals that have already had a chance to develop personalities and habits, both good and bad, then this may not apply, at least not until you have a new generation to work with. But if you're lucky enough to be starting with goat kids, you'll have the opportunity in their first months to shape them, both in their relationship to the fence and in their relationship to you.

Karl and I believe in positive reinforcement, and from the beginning, we've established routines and rewarded our goats with treats (a small handful of grain or chunks of pressed alfalfa) when they follow our commands. We're not teaching them to do tricks—though we did learn this method of behavior modification from Ringling Brothers' goat trainer—but we are teaching them the way to live on our

Vern Charles feeds mini marshmallows to a La Mancha dairy goat at Charles Ranch in Delta, Colorado. Finding the right treat can be a valuable tool for training and controlling your goats. Some farmers use pressed alfalfa or the promise of a handful of grain.

farm. Since part of our property remains unfenced, we lead the goats down to the paddock every morning with a small cup of grain and bring them back to the barn again in the evening when it's time for milking. We began leading them on leashes and actually feeding them their morning grain in the paddock as soon as we brought them home. We gradually dispensed with the leashes and led them with the grain cup, usually into the paddock, but sometimes on walks through our woods. Now, they'll simply follow us, grain or no. We find that it's easier to lure than to drag, and we know other farmers who use everything from alfalfa to mini-marshmallows to keep their goats in line.

This digression applies to fencing because if and when your fencing fails, you'll want to be able to get your goats under control as quickly as possible. This may mean enticing them with grain until you're close enough to grab their collar, or if they're very agitated, it may mean speaking calmly to them while you slowly approach, then petting and soothing them until they calm down. In any scenario, you want to know your goats' likes and dislikes, and you want your goats to know and trust you.

There are a lot of books that recommend using your goats' horns (if they have them) as handles, and certainly, as a last resort, that's a convenient option. We've found, though, that if we touch our goats' horns, they begin to think it's okay to touch *us* with their horns, and that becomes simply dangerous. I've found that when I avoid grabbing my goats by the horns and instead use a collar or lead, I feel more in control of the situation, no matter how frantic and frazzled I am by their escape.

FENCING SYSTEMS

With goats, as with any livestock, fencing should be thought of in terms of a system of containment, rather than simply as a string of posts and wire. The way you choose to contain your goats is a big piece of your overall management plan. The more effective your fences are, the more relaxed you can be about your goats, both in terms of predators and in terms of escape. The more land your goats have to explore, the less bored they will be, and the less likely they'll be to make mischief. And the more ground they cover—either roaming free on a large piece of property or in a mobile pen that is regularly moved—the less vulnerable they'll be to ailments associated with dense animal concentration, like parasites and scours.

Ideally, goats of any type are contained within a few layers of fencing: 1) a perimeter fence that marks your property's boundary and is the first defense against predators, 2) an inner fence that marks the goats' fields or paddocks, and, if applicable, 3) temporary fencing that subdivides the goats' area. Especially with dairy goats, you may also want to fence a corral or yard to bring the animals close to the barn at milking time. It's best if your barn opens directly into the fenced fields. Depending upon the size of your herd, however, it's also possible to lead the goats from the barn to the pasture.

PERIMETER FENCING

This chapter is a classic case of "do as I say, not as I do." As I mentioned earlier, when Karl and I got our goats, our property was entirely unfenced, except for a line of white pickets along the road in front of the house. We have gradually begun fencing the three open acres on which we keep our goats, which were marked when we bought the house by wooden posts in various states of decay. We generally keep a few rolls of four-foot wire fencing in the barn, and when we have a spare moment, we string the fence between posts until we come to one that's too rotten and must be replaced. Then we abandon the project until we think of it again, usually a few months later. Where we've strung the fencing, the wire is not stretched taut, it's not electrified, and any self-respecting goat could go through it in a second. We depend entirely upon our interior fencing for goat containment, and upon our goats' training and good nature when they escape. This is not a system I would recommend. We have only one line of defense against predators, and if the electrified goat fence shorts out or is somehow compromised, our herd could easily decide to go exploring.

Because we live in a relatively populated area, the goats are in the barn at night, and I'm home most of the day to keep an eye on the critters, we're able to take this laissez-faire attitude toward fencing. Except for the occasional bear sighting or nighttime coyote song, our predator concerns are mostly stray dogs, and if we're going to be gone for more than a few hours, we simply put the animals in the barn and close the door. With a small herd, it's possible to keep your goats this way, but it's certainly led to stressful situations for me, including some memorable waddling as I tried to shoo the goats out of the garden during my last month of pregnancy. If a little preparation and fencing forethought can spare you such indignities, I would highly recommend it.

A La Mancha dairy goat kid sticks its head through a welded wire panel at the Charles Ranch in Delta, Colorado. The stiff and sturdy panels can be used to reinforce other types of livestock pens or gates with large gaps to prevent smaller goats or kids from slipping through.

In general, exterior perimeter fencing is made of woven wire with one or two strands of tautly stretched barbed or electrified wire at the top and bottom. This is not to be confused with welded wire, which is not as strong and is generally made of a smaller gauge wire. Welded wire will keep out stray dogs, but determined predators and curious goats will have no trouble going through it.

Woven wire fences are made of horizontal lines of smooth wire connected by vertical wires called "stays." Most manufacturers construct them of either nine or twelve and a half gauge wire, and for goats, the heavier is recommended. Generally, the space between the wires expands as the fence gets taller, so the openings in the fence are larger at the top. A four-foot woven wire fence, fortified with barbed

Tippy peeks under a barn door at Frank and Carol Webber's Highland Badger Farm in Ellison Bay, Wisconsin. Goats' natural curiosity is one of the things that makes them so hard to contain.

wire or electrified high-tensile wire, is usually considered a good fencing option for goats, both as perimeter and interior fences.

An alternative to buying rolled fencing is the purchase of fence panels, generally made of heavy-gauge welded wire. These vary in height and can be purchased in either eight- or sixteen-foot lengths. Panels are sturdy and resist sagging, but they aren't flexible like rolled fencing, so they work best as interior fences on relatively flat ground. Once installed, they can be treated the same way as woven wire fencing.

Barbed wire is sometimes recommended as a perimeter fence, especially when the animals have a large piece of property to wander. Barbed wire is more economical than woven wire, though when calculating the cost, you should keep in mind that you'll need approximately eight horizontal strands and at least a couple of vertical wire "stays" between posts. Fifteen-gauge wire is generally recommended, and it should never be electrified. The drawbacks of barbed wire are its potential to cause injury to the goats and its uneven protection against determined predators. That said, if it's supplemented with an electrified interior fence, a barbed wire fence makes a good boundary fence and is certainly a deterrent to both predators and escape artists.

Board fences, like those used on horse farms, are generally not recommended for goats, since the gaps between the boards are often wide enough for a wily goat to slip through. They can be expensive to construct and maintain, too. Because they are aesthetically pleasing, however, we've seen some farmers run a piece of board along the outside at the top of a woven wire fence. If you have the means and the inclination, this makes a strong and attractive fence.

Percival, a two-year-old Alpine wether, gazes past the perimeter of the portable electric fencing at Ten Apple Farm in southern Maine. This plastic and wire fence contains goats and allows their owners to move them around for fresh browse (saplings, weeds, and tough vegetation). If you keep the fence clear of high weeds and train your goats on it when they are young, they quickly learn to respect the 6,000-volt charge.

Gates, for any of these fences, should be as strong as possible, with narrow openings and sturdy hinges and locks. There are a variety of gates that will keep goats contained, but the most popular is probably the bar gate, made of open horizontal bars spaced at relatively narrow intervals. One of this gate's many advantages is that the open bars make it difficult for goats to get traction and push against the gate. Choose latches that can be opened with one hand, since you'll usually be carrying something—or leading a goat—when going through. As with all goat containment, when choosing a gate, look for the sturdiest material and construction you can find.

Good fences, as the saying goes, make good neighbors. It's never a bad idea to mark off your property, and if you're considering buying animals, the sooner you do it, the better. When you realize that you need a perimeter fence, it will probably be too late to do you much good. Whatever you choose as your boundary, if you can bring your goats home to a piece of property that's already been fenced, you'll save yourself time and stress later on.

INNER AND TEMPORARY FENCING

Inner fencing is the layer of containment with which your goats will have the most direct contact. If you use electric fencing, your goats will learn very quickly to respect the fence. A few zaps after curious nose nudges will teach them that

the faint buzz and click they hear means "stay away." For electrified fencing, high-tensile, smooth wire is the cheapest and most reliable option. High-tensile wires are desirable because they can be stretched tightly without breaking and will resist sagging over time, but they need strong posts and corners to do the job. If you're using exclusively high-tensile wire for your goat fence (and not a combination of high-tensile strands and woven wire), at least five strands of twelve-gauge wire are generally recommended. Either a battery powered energizer or a solar charger is needed to electrify the fence, and for goats, a pulse with at least a 4,000-volt charge is recommended. Some chargers are now designed to be "low impedance," meaning that their charge isn't drained by vegetation or other materials touching the fence. Whether or not your charger is low impedance, your goats will learn after their first few shocks to avoid the click that signals that the fence is on.

Portable electrified netting is another option for fencing, and we find it ideal for moving goats around our property. Rather than permanent posts and high-tensile wire, we use a mobile net fence that attaches to portable plastic posts that are staked into the ground. Every few days, we move the pen and the goats can browse in a new spot, giving the rest of the field a chance to recover. While it's a more expensive fence, we like mobile netting because it's quick to erect and it allows us to easily move and change the shape of the pen. When we want to concentrate the goats in a certain area—the brushy entrance to the orchard, for instance—we simply move the fence where we want it. In the winter, when the goats spend most of their time in the barn, we can roll up the fence for storage and pull it out only when we want to give the goats some fresh air. There are certainly drawbacks to portable fencing—its relatively short length means that we must constantly move it to give the goats fresh vegetation—but we find that on our property, the pros outweigh the cons.

With all electrified fencing, it's imperative that you follow the manufacturer's instructions, and if your supplier recommends other products to use with the fence and energizer, I would suggest buying them. The $15 fence tester may seem like a frivolous accessory when you're tallying up your first mammoth fencing order, but it will make a world of difference later on when you're trying to figure out why your goats keep escaping.

SUPPLEMENTAL PROTECTION

While fencing can do a lot to keep your goats safe, you may also want to consider adding a guard animal to your herd. Large dogs, like Maremmas, Anatolian Shepherds, and Great Pyrenees, are traditionally used as herd dogs with goats, and have been bred over centuries to be effective guard animals. Training and integrating a guard dog into your herd is a subject too vast for the scope of this book, and it's too important to be treated lightly. If you are considering bringing a working dog into your herd, make sure that you do thorough research beforehand, and as with any animal, try to acquire the dog from someone who can remain a resource and mentor.

Other popular guard animals for goats include llamas, which can both sound the alarm and physically defend the herd, and donkeys, which can alert you to the presence of an intruder. Like goats, any guard animals you bring home will need to be trained and maintained, and before you commit to one, make sure that you have the energy and resources to care for it. Karl and I know from our unruly house pets that we don't have the time or discipline to train a guard dog. Though I dream of someday adding a donkey, at this point our herd is on its own, which in our situation has been fine.

Your decisions about guard animals will be influenced by many factors, including the predators in your area and your own interests, and you may decide to wait until you've learned to live with goats before bringing another animal into the herd. At our farm, the philosophy on animal acquisition is if there's any hesitation, it's more prudent to wait. If, on the other hand, you're eager to put in the hours to train a working dog, I would encourage you to do so—they're marvels to watch in the field.

PROTECTING YOUR GARDEN: GUARDING AGAINST THE INEVITABLE

As comprehensive and sturdy as your fencing may be, it is inevitable that at some point your goats will make their great escape. Whether it's a full moon that inspires them or a predator that spooks them, your goats will at one time or another find their way to the other side of the fence. When this happens, they will go where their little hearts desire: most likely into the garden or orchard that's been tempting them from afar.

If you have a large kitchen garden, like we do, or a few saplings or fruit trees, your goats can inflict a lot of damage in a very short amount of time. They'll nibble here and there, helping themselves to the tender florets on your broccoli and the juicy greens of your beets, tearing off strips of bark from your favorite pears and pines. With shocking speed, they'll reduce your carefully planned garden to a few scattered stems and stumps.

The good news is that, at least in theory, it's relatively easy to avoid this grim scene. By erecting strong fencing around your garden—fence panels work well, or the same woven wire that you may be using around your paddock, with a snug-fitting, locked gate—you should be able to keep out curious goats. To protect your trees, prune them so that no low hanging branches dangle at goat-height. Wrapping trunks with either wire mesh or a close-fitting "cage" of wooden slats is an effective protection for mature trees. For saplings, the best protection is fencing, as it is for the garden.

CHAPTER FOUR

Feeding Your Goats

When Karl and I first began thinking about goats, we quickly learned that there are as many theories about goat care as there are goat owners. This is especially true in the area of goat feed, where a nutritional advantage can result in steadier growth, increased milk production, and, while a doe is pregnant, even an increase in the number of kids she ultimately delivers. With so much at stake, goat farmers are often passionate about the choices they make with regard to feed. As you get deeper into your life with goats, you will probably form a preference for a certain balance in your feed rations. At the beginning, however, it's important to master the essentials of goat nutrition.

First and foremost: Contrary to popular wisdom, goats are not tin-can eating omnivores. They will not "eat anything," and taking a cavalier attitude toward their diet can result in serious but easily avoidable problems. Even the most seasoned farmers encounter accidents with their feed, and illness that results from nutritional deficiencies (or excesses) will be addressed in greater detail in chapter seven. In this chapter, I'll focus on trying to ensure that your goats get the foods they need.

Flyrod stands on hind legs to munch on oak leaves at Ten Apple Farm in southern Maine. Goats will eat leaves and strip the bark from trees, so valued saplings should have extra protection if they're near the goat pen.

Goats are ruminants, in the same family as cows, sheep, and deer. More specifically, they are browsers. This means that, like the deer that wander through

 A Kiko buck and several Kiko does look for grain in the tall grass and clover at Ruble Conatser's Mountain Goat Ranch in Jamestown, Tennessee. Grain is a necessary supplement for lactating does and a good source of nutrients that might not be found in local browse or hay.

our orchard and nibble the tips off low branches, goats prefer weeds, woody stems, and bark to grassy pasture. Because their diet is naturally varied, goats are designed to be choosy when it comes to their food, and they can actually be quite finicky when they aren't offered their preferred foods. This can make it a challenge to feed your goats an adequately balanced diet, or it can make it pretty simple. The spectrum of choices for management is wide: You can intensively control your goats' feed rations, choosing a combination of proteins, roughage, and minerals and administering it in exact amounts. You can allow your goats to browse freely and choose their own food, turning them loose on property that's dense with brushy vegetation. Or you can use some combination of the two methods, allowing them free choice with some supplements. The choices are yours, and will be influenced by many factors, including your climate, acreage, and herd size, as well as your own temperament and interests. Once you understand the fundamentals of your goats' digestive system and nutritional needs, you can choose a feeding system that works for you. Until then, here are some basics.

WHAT IT MEANS TO BE A RUMINANT

Ruminants are hoofed, generally horned quadrupeds that chew their cud. The suborder *Ruminantia* includes goats, sheep, cattle, deer, and even giraffes, and is defined by the fact that the animals' four-chambered stomachs allow cellulose to be broken down, regurgitated, fermented, and finally turned into digestible nutrients. Ruminants can quickly ingest large quantities of food, swallow it nearly whole, and bring it up again to chew later, an adaptation that may have originated to protect these gentle grass-eating creatures from predators. They are vegetarians, and through the act of ruminating, their bodies create their own protein during digestion.

Digestion occurs in the ruminant's four-chambered stomach as food travels from the rumen to the reticulum, then to the omasum, and finally to the abomasum or rennet stomach, where enzymes and hydrochloric acid are secreted. The rumen makes up the largest portion of the stomach, accommodating between three and six gallons of liquids and solids in an adult goat, and it's there that the food is broken down and fermented. When an animal is chewing its cud, it is regurgitating food from the rumen and then breaking it down and mixing it with saliva. On average, goats move their jaws at least forty thousand times a day as they chew and re-chew their food. Like all organs, the rumen is made of muscle, which contracts to bring food up, to mix it, and to squeeze it further into the digestive system. Interestingly, ruminants are not born with an active rumen; the milk they consume while nursing passes directly from the esophagus to the abomasum or rennet stomach. It is only when they begin to consume roughage that the rumen expands and starts to function.

I won't go into much further scientific detail about digestion, but the key thing to remember about ruminants in general and your goats specifically is that they are designed to exist on grasses and browse (saplings, weeds, and tough vegetation). Their digestive systems will be healthiest when their rumen gets a good

workout, which means that whatever else you may use to supplement it, you must
be sure to give your goats a whole lot of roughage. The three factors that will create
a healthy rumen are as follows.

Roughage

Coarse, fibrous feed—like hay and weedy browse—will keep the rumen working
away, churning and fermenting as it breaks down the cellulose in the plants.
Unlike tender new shoots, which contain energy but not much fiber, roughage
must be chewed and re-chewed, stimulating production of saliva, which helps
lubricate the entire digestive system. And in order for a goat's body to access the
nutrients in roughage, it must be broken down by the rumen, which keeps the

Lani Malmberg's herd of weed-eating Cashmere goats moves across a field in Cheyenne,
Wyoming, eating as they go. Natural browsers, goats eat tall woody brush and weeds first,
leaving grass as their last choice. This makes them ideal for clearing weeds and invasive
plants. They fertilize and aerate the soil with their hooves as they go.

organ active and keeps the entire digestive system chugging along. It is possible for feed to be too coarse—bark and woody browse can sometimes be only partially broken down by the rumen—but in general, the stalks and mature leaves of most plants make a fine diet for goats.

Feed with a minimum of sugars and starches

Goats, like humans, are attracted to a diet of simple carbohydrates. Grain that is fortified with minerals and sweetened with molasses is a common goat feed, and when used in moderation does a wonderful job of supplementing nutrients for kids and pregnant and lactating does. We feed our goats coarse grain at each milking, and we use it as a bribe with our wethers, but a diet of too much sweet grain is the caprine equivalent of fast food: cheap and plentiful calories, but ultimately not what the body needs.

Consistent rumen acidity

The rumen is naturally slightly acidic, and the microorganisms necessary for proper digestion function best in this environment. In the wild, goats can manipulate their diet to keep their rumen's pH steady, choosing browse that is the animal equivalent of Alka Seltzer. In the barn, however, it's important for you to keep an eye on your goat's belly, watching for signs of distension, and make sure that their diet includes plenty of roughage. Too much grain will cause the rumen to become acidic, and if left unaddressed can lead to serious digestive problems like bloat.

MEETING YOUR GOATS' NUTRITIONAL NEEDS

I can't stress emphatically enough that a well-functioning rumen is one of the keys to a healthy and happy goat. In the wild, goats subsist entirely on vegetation of their own choosing, and while it can be very beneficial to supplement the diet of domesticated animals with extra nutrients, minerals, and vitamins, it's also possible to overdo it. As goatherds, Karl and I are inclined to keep things as close to natural as possible, which means that if there's something on our property that our goats seem eager to eat, we usually let them, unless it's poisonous or growing in the gar-

den. To make sure that our
goats are getting adequate
nutrition year-round, however, we have a few rules
of thumb.

WATER

At our farm, the first rule
for all of the animals—
humans included—is that
they have an abundance
of fresh, clean water. It's
amazing how quickly dehy-
dration can become a problem, even in a temperate

 Alpine and Saanen dairy goats drink from a tub of water
at Redwood Hill Dairy in Sebastopol, California.
Making sure your goats have plenty of fresh, clean water
is extremely important, especially for lactating does.

climate like Maine, if your animals don't have plenty to drink. In a hotter or more
humid climate, I would recommend an automatic watering system, if it's within
your budget. In our environment, it suffices to fill a clean bucket with water at
each feeding, and to check and top it off periodically during the day. It should go
without saying that there be no poop or slime in the bucket.

In the winter, when things are likely to freeze, we fill the buckets with warm
water, and simply tap out the ice or thaw the buckets in the house when necessary.
You can buy all kinds of equipment to keep the buckets from freezing, and if you
live in a cold climate and you're gone during the day, I would recommend looking
into an electric water heater.

The essential thing is that your goats
have constant access to good, clean water.
Hydration will keep your goats' digestive and
urinary tracts functioning properly, will ease
the stress on the mammary systems of your
lactating does, and will help keep their
entire bodies in tip-top shape. In short,
water is a wonderful thing.

 An Alpine dairy goat drinks
water from a special spout at
Rick and Lora Lea Misterly's
Quillisascut Farm in Rice,
Washington. "Goats are very
persistent. They never give
up," says Lora Lea.

Above and Opposite: A pair of Boer goats eats from a hay feeder at Ken and Janice Spaulding's Stony Knolls farm in Saint Albans, Maine. The feeder is Ken's own design and opens with hinges on one side so a large round bale can be rolled in. Ken warns to watch out after introducing a newly built wooden feeder as goats are prone to get abscesses from splinters.

HAY

The second rule we follow is to keep the hay feeders full. If you can make your own hay, by all means do it—you'll know exactly what you're getting and if you already have the equipment, it will be cheaper to hay your own fields than to buy from someone else. If, like ours, your property doesn't have enough open fields to make hay, develop a relationship with a hay supplier. If you don't know a supplier in your area yet, you can generally find one by checking with your extension office or asking at your local farm store. Our extension office maintains the online Maine Hay Directory, which lists contact information for sellers by county, as well as some links to sites that offer advice about sampling, testing, and buying hay. Other state extension offices provide the same service. We found our own hay supplier through our local farm store, which keeps the phone numbers of a few area hay-makers behind the counter. However you find them, make sure that you're buying hay from someone you trust, who gives you a fair price and makes it as easy as possible for you to get your hay from their fields into your storage.

As far as composition of hay and its nutrient makeup go, the specifics make a broad topic, and one on which many farmers are passionate. We take a relatively open-minded approach: As long as the hay is in good condition and our goats like it, we feed it to them. The only way to truly know what you're getting is to have a sample analyzed, and so far, we haven't done that. In general, young goats and pregnant and lactating does prefer legume hays such as alfalfa, clover, lespedeza, soybean, and vetch, while adult males and non-lactating does do well on mixed grass and legume hay. Unless you're intensively managing your goats' feed, however, you can usually give all of your goats the same hay.

Here, we feed our goats second crop, mixed grass hay. Second crop simply means that we buy hay that's made when our supplier does her second cutting, usually in midsummer. (First cut hay is longer, and can be difficult for goats to get into their small mouths.) Our supplier sells mostly to people with horses, and in general, good horse hay will work for goats, too. When you're buying hay, let your supplier know that it's for goats and she should steer you to the right stuff. The hay should be green and sweet smelling, and if you open up a bale it should

An Alpine dairy goat gets its head stuck in a bucket while looking for food at the American Dairy Goat Association's 2004 National Show in Harrisburg, Pennsylvania.

easily come apart into thick "flakes," or chunks. Make sure that the bales are dry and properly cured before you buy them—goats will refuse any hay that's moldy or dusty.

When you're buying hay, you can usually arrange to buy it in the field, on the farm, or to have it delivered. If you buy it in the field, you will often get a better price because you're saving your supplier the labor of loading it onto their hay rack and the transportation cost of taking it out of the field. The drawback is that you need to be "on-call" to pick up the hay. Since making hay is weather-dependent (you know the saying, "make hay while the sun shines"), you need to make sure you'll be available to come get the hay when it's ready. Buying hay at your supplier's farm is often the most convenient plan, and if storage at your farm is tight, you can sometimes arrange to pay in advance and pick up your bales as you need them. Having your hay delivered is generally the most expensive option (our hay supplier charges 25 cents more per bale if it's delivered), but if you don't have the equipment to transport your hay, it's sometimes the only option. Before you decide how you'll buy your hay, make sure to discuss all the options with your supplier.

One last important thing to mention about hay is storage. For our goats, we plan on about one bale per goat per week in the winter. (This equals roughly four

and a half pounds of hay per goat per day.) We keep four goats year-round, so when we buy our hay in the summer, we try to get at least 150 bales in the loft, which we stack and cover with tarps to keep free of dust, light, and moisture. Prolonged exposure to sunlight will discolor hay (one reason you want to buy bales that are a rich green), and it will also begin the process of decomposition. When choosing a spot for storage, try to avoid direct light. You also want your hay to be stored off the ground. Hay lofts are ideal because they allow for plenty

GROWING A GOAT GARDEN

Though you generally work hard to keep your goats out of the garden, it can be a fun project to mark off a plot just for them. We found plans for a goat garden in an issue of the *Capri-Zette*, the newsletter of the Nebraska Dairy Goat Industry, which had adapted a passage from J. E. Eberhardt's classic *Good Beginnings with Dairy Goats*. The book includes a detailed listing of nutritional information for each plant, and recommends planting:

Carrots—feed goats greens and chopped roots

Pumpkins and winter squash—feed goats protein-rich seeds and chopped fruit

Sunflowers—goats will eat entire plant, but wait until it develops a seed-head to harvest

Comfrey—feed leaves and stalks either fresh or dried

Kale—in moderation, feed leaves and stems

Rugosa roses—feed goats dried rose-hips from this Asian variety

Apples—in moderation, feed goats both fallen fruit and pruned branches

of air circulation, but even putting the bales up on pallets will do. You want to avoid any chance that the hay will get damp from below. When putting hay up for storage, it's important that the bales are completely dry—any bales that are heavy when you buy them should be suspect, and you can test them for moisture by simply sticking your fingers in to see if the hay feels wet and hot. Damp hay will mold over time, but it's also a serious fire hazard, since it can spontaneously combust. A good hay supplier won't sell you damp hay, and if yours does, it's time to find someone new.

Cultivating a good relationship with your hay supplier is helpful when buying hay; choosing a solidly designed hay feeder is important when giving that hay to your goats. There are many different designs out there, from keyhole feeders especially made for goats to metal racks created for horses and cattle. In our experience, there are no perfect feeders, and goats are notorious for wasting hay.

Our preference is for feeders that keep the hay off the ground and are designed with a small enough space between the slats that there's some resistance so the goats have to work for their hay. As with most equipment, your choice of feeder will depend a lot upon your own situation. Some mount on the wall, some hook over the side of the pen, and your decision will be influenced by the design of your pen. Keep in mind that as a general rule, goats don't like to eat hay directly off the ground, so you'll want something with a base that's at least a little elevated. Also, since they tend to spread hay around pretty lavishly, you may want to consider putting a tray or trough directly underneath the feeder. Here, we have a raised tray with a two-inch lip under our hay feeder to catch stray bits. It's only moderately effective, but it does save us a little from each bale, which adds up over the course of the winter.

SUPPLEMENTAL FEED

While the core of your goats' feed will be made up of roughage, most goat farmers also choose to include supplemental minerals and some grain in their animals' diet. This is another area of feed in which many farmers become passionate. Experienced goatherds and those with a background in animal science will often create their own balance of nutrients, and if you have the inclination and a comprehensive understanding of the nutritional needs of small ruminants (much more than is discussed in this book), that's certainly the option that gives you the most control over your animals' diet. It takes a lot of work, however, and miscalculation can have drastic consequences to the health of your herd. Though it may initially seem cheaper to mix your own feed, I would caution against doing it simply for financial reasons, and recommend it only if you feel that the premixed feed available in your area isn't up to snuff. If you're considering mixing your own feed, you'll find some resources in Appendix C of this book, but I would highly recommend consulting your local extension agent before proceeding.

For those who are new to goats, it's an easier and safer choice to buy a premixed feed and talk with your local extension officer about the minerals that may need to be supplemented in your goats' diet. Mineral content varies among soils, and what's available in the vegetation your goats are consuming will depend upon your location. In some parts of the country, like New England, there's a serious deficiency of the trace mineral selenium in the soil, and the standard premixed goat feed in our area includes a supplement. In other parts of the country, however, an excess of selenium in the soil is a problem, and adding a supplement to what goats naturally consume in their browse could be toxic. As with all questions that are specific to your area, it's best to consult with an extension agent if you have any concerns.

Once you've decided on the type of grain to use, you'll want to use it judiciously. Obesity can be a real problem in goats, and can lead to serious illness, from urinary calculi in wethers to pregnancy complications in does. While feeding your goats grain, you should regularly assess their body condition to make sure that they're neither too fat nor too thin.

We give our goats grain twice a day, feeding our lactating does approximately half a pound of grain for every pound of milk they're producing. Dry, pregnant does get about half a pound of grain each day to boost their caloric intake. In their first few months, kids get about the same amount, though we leave it in a grain pan at first so they have "free choice," and they gradually work up to a pound of grain each day, divided between their two feedings. Our wethers don't actually need any grain, but we still give them some—no more than a quarter pound a day—as a treat.

Many farmers leave mineralized salt, also called salt licks or salt blocks, accessible to their goats all the time, either outside in their paddock or hanging in their pen. This works well, as it allows the animals to self-regulate when they feel a deficiency in their diet.

TREATS

Everyone loves a treat, and goats are no exception. We've heard of virtually everything being used as goat treats, from animal crackers to pressed alfalfa. In general, we try to avoid heavily processed foods and give our goats a handful of grain, a little pressed alfalfa (sold in two-ounce cubes that are intended for rabbits), or leftover bits from the garden when they deserve a reward—or need a lure. As with anything, your goats will have their own preferences and it's best to experiment a little to see what they like. Ours love cabbage and beet greens, for example, but have little use for kale until it's the last thing growing in the garden. After a few tries, you'll know what yours like. It will make a world of difference when you need them to hold still for a hoof trimming or a vaccine, or when you just need a little nuzzle and some goat love.

CHAPTER FIVE

On Bucks and Breeding

Whether you want to keep a dairy doe lactating, to generate kids for sale, or simply to grow your herd from within, at some point you will probably want to breed your goats. There are several ways to go about doing this, and each method of breeding has benefits and drawbacks. You can opt to inseminate your doe in a short, managed breeding session with someone else's buck, you can artificially inseminate the doe, or you can keep a buck yourself. If you are considering including a buck in your herd, it will certainly make the act of breeding easier, if not inevitable, but it will also present its own set of challenges, which should be considered at the outset.

Simply put, keeping a buck can be a lot of work. In addition to general goat needs like food, shelter, and grooming, bucks bring with them the added challenge of testosterone. This single hormone additive makes bucks exponentially more difficult to manage. They are fearless, often testing their boundaries, clashing with each other, and barreling through their fences. They are assertive, sometimes becoming dangerously aggressive while in rut, the annual period of sexual arousal, which can last several months. And they are stinky, attracting does by enurination—spraying their beards with their own urine—and by producing that distinctively goaty smell from their scent glands. Though they are impressively free of inhibitions and their antics can be entertaining to watch, for the new or small-scale

A Toggenburg buck pauses on his way out to pasture on a sunny October day at Pleasant Valley Acres, Cumberland, Maine.

Lightning, a Kiko buck and herd sire, peers over the fence at the Mountain Goat Ranch in Jamestown, Tennessee.

farmer, keeping a buck may prove more troublesome than beneficial.

On the positive side, however, if you have a buck, when you want to breed you'll be able to watch your does for the telltale signs of estrus, put them in with the buck when their cycle peaks, and let nature take its course. Or, if you're less concerned with the exact date of conception and simply want to track who's breeding with whom, you can put a few does in with the buck for a month, eliminating the need for an estrus vigil. It's a good system, especially if you have a large enough herd that breeding logistics and fees are daunting. The drawback is that, for this month or two of work, you need to keep your buck happy, healthy, and contained for the entire year.

Before I go any further, I should mention that we don't keep a buck, but we do breed our goats yearly. There are several wonderful Alpine breeders in our area, people with healthy herds, whose philosophies of animal care we agree with, and each fall we watch our girls for signs of estrus, load them up, and drive them over to be bred, for a small fee.

I'll go into more detail about the specifics of breeding later in this chapter, but I mention the process here to contrast our annual breeding sessions with the year-round care that keeping a buck will require. I am not against keeping bucks, even in a small herd, and there have been bucklings born on our farm that I've been loathe to give up, but including a buck in your herd will add a different level of management to your life with goats. Many have steady, mellow, gentlemanly temperaments, but if your buck is frisky, adding him to the herd can also add stress.

KEEPING BUCKS

Puck, Pan, the Satyrs: The mythology of these priapic goat-men is one of licentiousness and ribaldry, of creatures led by their loins into all kinds of mischief. As those who keep bucks will be quick to tell you, this cliché isn't all that far off the mark. The wonderful things about bucks are the same things that can make them difficult. They are big, agile beasts that take great pleasure in their physicality. Their total lack of restraint can be amazing to watch, but it can also make them very difficult to control.

A few words about the anatomy of a buck: They are very large animals, most weighing at least 200 pounds, with some male Boers topping the scales at 350 pounds. Unlike castrated males, bucks have developed musky scent glands both at the crown of their head, near their horns (if they have them), and in their hind legs, near the base of their tail. They are also "intact," which means that they still

Sue Buntin of Marion, Kentucky, waits to show a Boer buck at the International Goat Days Family Festival in Millington, Tennessee. Sue and Wade Buntin are big supporters of 4-H programs, and they serve as the adult leaders of the Crittenden County Goat Getters 4-H Club.

have their scrotum, an oval sack that ideally contains two large symmetrical testicles. Depending upon their breed and use, bucks may retain their horns, which can range from a set of Boer's close, sickle-shaped arcs to a Kiko's wide-spread spirals. Like does and wethers, dairy bucks are generally disbudded soon after birth, and must be if you plan to show them with the American Dairy Goat Association.

Like all creatures, the temperament of bucks varies widely, and their personalities are influenced by a variety of factors. These include your interaction with them, the comfort of their living situation, and their fundamental make-up. If you spend a lot of time with your buck, bottle-raising him and working with him from his infancy, he may be relatively docile for most of the year. (The key thing to remember is "most of the year"—the surge of hormones in breeding season changes everything.) If you don't spend much time with your buck, if he was "dam raised" (nursed by his mother), or if he's simply got an unpleasant streak, a buck can seem more wild than domesticated, especially in comparison to the rest of your herd.

 Chief, a Saanen dairy goat buck and herd sire at Haystack Mountain Dairy in Boulder, Colorado, smiles for the camera. Owner and award-winning cheese maker, Jim Schott, says the thing he loves about goats is that they are "humanlike."

No matter how sweet and gentle he may be the rest of the year, when your buck comes into rut, he might not be himself. When he's ready to breed, he should be handled with extreme care, especially if he has horns. Your own buck may be even tempered, but until you're certain of this, it doesn't hurt to be cautious. This means taking some of the following precautions that may seem extreme the rest of the year:

- Keep your wits about you. Breeding season is not the time to daydream in the barn or to listen to talk radio while doing chores. Making sure that you're aware of your environment is essential to staying in control of your goats when they're feeling frisky.

- Don't turn your back on your buck. This is generally a sound practice, but during breeding season, it's an especially good idea. If your buck decides to be aggressive, he'll eye you, size you up, and give warning signs before he makes a rush. Make sure you've got your eye on your buck any time you're in the pen with him.

- Never let a buck get between you and your only means of exit. Any time you have to go into the pen with your buck, make sure that you can make a quick escape. Keep your face to the buck and your body as close as you can to the fence, while also accomplishing the chore that brought you into the pen.

- If you are female, be especially careful around bucks in rut. Goats often think of their keepers as members of the herd, and bucks in rut may not distinguish between a fetching doe and their female owner. If your buck starts to rub against you and blubber, get out of the pen!

- If your buck seems especially wound up and you cannot avoid going into the pen with him, take some kind of deterrent with you. We've heard of goat farmers running the gamut of weaponry from broomsticks to Tasers, but if possible, I would recommend something like pepper spray, which will discourage a goat without harming him. I've also heard of producers carrying a squirt-bottle of vinegar mixed with water, but if you're really concerned, you might want something a little more forceful. In a dire situation, if you cannot escape an aggressive buck, we've read that grabbing the goat by his beard and maintaining a firm grasp until you've reached an exit is an effective strategy.

Of course, the best way to keep safe around a rutting buck is to simply stay out of his way, while keeping him contained. This last part is easier said than done, and we don't know a farmer who keeps bucks who hasn't had a rutting buck escape. Generally, he'll make his way to the does, and many a breeding program has been sabotaged by just a few feet of downed fencing. To minimize the chance

of this, most farmers keep their bucks in pens at some distance from the does, with a couple of fences between them.

If you're raising dairy goats, another reason to separate the bucks from the does is that the presence of bucks can sometimes cause an off-flavor in milk. Hormonal changes

 Jimmy Young of Wellington, Ohio, transfers straws of goat semen to a semen tank at the American Dairy Goat Association's 2003 National Convention in Nashville, Tennessee. Events like the ADGA National Convention are excellent opportunities to purchase or trade buck semen in order to improve and diversify the genetics of your herd.

caused by the proximity of bucks can give milk an extra goaty taste, and to avoid that, many small-scale dairy producers (like ourselves) opt not to keep a buck. On our farm, we notice that the sweet milk that we can usually count on from our Alpines becomes progressively "goatier" if we have maturing bucklings in the barn. This may suit your needs, especially if you're making cheese that you'd like to give a little pungent bite, but it certainly changes the flavor of your morning coffee!

The Economics of Keeping a Buck

From an economic perspective, keeping a buck is a very expensive way to acquire semen. At this writing, straws (plastic ampoules) of good quality semen for artificial insemination from registered purebred bucks of most breeds can be bought for less than $50, and scheduling an annual morning breeding session for a single doe is comparable in cost. (Of course, straws from champion bucks go for much, much more, but if you're in the market for that caliber of semen, you're probably already well-educated about the economics of breeding!) The maintenance of your buck—hay and grain, barn upkeep, veterinary visits, and secure fencing—will cost much more per year. That said, if you're planning to breed more than a couple of does, the cost of impregnation can certainly mount. While it's generally not a good idea to breed a buck with its offspring (unless you're linebreeding, which I'll discuss later in this chapter), with good management and careful record-keeping you can maintain a large number of does, and even grow the herd, with just two bucks. If you have an exceptional buck, his semen may also be in demand, which would put you on the receiving end of breeding fees.

Ultimately, when making the decision to keep a buck, you need to evaluate your own needs, the energy you're willing to spend on management, and the experience you're hoping to have with goats.

Breeding Your Goats

As I touched on earlier, if you're planning to breed your goats, there are several ways you can go about the physical impregnation (see Insemination page 72). In terms of goat management, you also have alternatives, and there are choices you'll need to make about breeding before you ever pull out the "buck rag" to kick-start estrus (see the mini glossary of breeding terms on page 64). I'll talk in some detail about management choices first, but for more detailed information, and any advice about what you might want to try with your herd, I would recommend consulting some of the resources mentioned in Appendix C and talking with an experienced goat keeper who can assess your herd. When you're new at raising

A MINI GLOSSARY OF BREEDING TERMS

If you've spent any time at goat shows or conferences, you've probably heard terms like *linebreeding, inbreeding,* and *heterosis.* You may also have heard words specific to the act of breeding like *rut* and *estrus, sperm motility, standing heat, settling,* and *freshening.* Here's a brief glossary of breeding terms to strip away the mystery and help you get started:

Abortion—termination of a pregnancy, often caused by injury, disease, or poor nutrition

Accelerated kidding—breeding a doe so that she kids more than once a year

Anestrus—period when a doe is not having estrous cycles, for most dairy goats, during the spring and summer

Artificial insemination (AI)—insertion of semen into a doe by artificial means

Breed—to impregnate a doe

Breeding season—time of year during which goats are bred, between early autumn and mid-winter for most breeds

Buck rag—piece of cloth that has been rubbed on a buck's scent gland, exposure to which will help spur estrus in a doe

Cloudburst—false pregnancy in which the uterus fills with fluid

Colostrum—thick, yellowish "pre-milk" secreted from udders immediately after birth, which is full of antibodies essential to development of kids

Conception—when an embryo is formed by the joining of male sperm and female ovum

Cover—the buck's mounting of the doe during breeding

Cross-breed—goat with parents of different breeds

Cull—selective removal of inferior animals from the herd

Dam—female parent

Estrous cycle—repeating series of physical and behavioral changes that occur from one period of estrus to another, in goats generally between eighteen and twenty-three days

Estrus—period directly preceding ovulation, during which a doe is receptive to bucks and can become pregnant, also called "heat"

First freshener—doe that is pregnant for the first time

Flagging—tail wagging by a doe to indicate estrus

Flehmen response—in bucks, pulling back the upper lip into a stiff grin to sniff for the presence of pheromones

Flushing—increase in nutrient intake in does in preparation for breeding

Freshening—period when a doe gives birth and begins to lactate

Gestation—period during which a kid develops in utero; in goats, 150 days

Heat—see *estrus*

Heat cycle—see *estrous cycle*

Heterosis—increased hardiness resulting from cross-breeding, also referred to as "hybrid vigor"

Inbreeding—intentional breeding of closely related goats, such as siblings or parent and offspring, to intensify certain desirable traits

In heat—period of estrus, in goats usually between twelve and thirty-six hours

In kid—pregnant

In milk—producing milk

Jug—small pen in which doe and kid are kept immediately after birth, not used with dairy goats. See also *mothering pen.*

Kidding—giving birth

Kidding pen—separate pen in which does will give birth

Lactating—producing milk

Linebreeding—intentional breeding of goats with a shared ancestor, to intensify certain desirable traits

Mothering pen—see *jug*

Open doe—doe that is not pregnant

Ovulation—release of an egg from the ovary

Progeny—offspring

Puberty—sexual maturity, reached by most does between four and eight months and most bucks between three and four months

Rut—period during which bucks are intensely interested in breeding, generally between early autumn and mid-winter

Settle—to conceive

Silent heat—estrus during which behavioral signs are minimal or nonexistent

Sire—male parent

Standing heat—period during estrus when a doe allows the buck to mount her

Sperm motility—movement of sperm

Sperm viability—whether sperm is able to fertilize an egg

Weanling—a kid from weaning until one year old

Yearling—male or female goat between one and two years old

goats (and even when you're not), there is simply nothing more valuable than a set of experienced eyes to help appraise your herd's strengths and places for improvement. By carefully choosing a sire, you may be able to add desirable traits to the next generations of goats on your farm.

When Karl and I first became interested in goats, we were confounded by many of the terms related to breeding, and the whole process seemed incredibly daunting. As with so much of animal management, in breeding your goats you have many decisions to make, and at the beginning they can all seem overwhelming. The most important thing to remember is that no decision is permanent; no matter what choices you make this time, most likely, you'll get another chance to breed next year.

Choosing a Sire

The greatest variable in breeding is choosing a partner for your doe, and there are a number of ways you can approach that decision. Since your goal is to improve the next generation of kids, you'll want to choose a sire that intensifies your doe's best traits and makes up for her deficiencies. For example, if your doe is petite, you'll want to breed her to a buck of large stature. You can also get more specific: If you have a milking doe with a sweet disposition and a sound body, but whose udder leaves something to be desired, you'll want to choose a buck that has been known to sire does with good mammary systems.

The more experienced you get with this kind of breeding, the more you may want to experiment. Linebreeding and inbreeding are techniques used by many farmers to amplify traits that are evident among goats that are related to one another. In each of these practices, does are bred to bucks to which they're closely related: Siblings to each other and parents to progeny are the most common pairings. Wonderful results can be achieved with linebreeding, but it can also intensify negative and recessive traits. You must be prepared to cull heavily if you're considering this kind of intensively managed breeding.

At the other end of the management spectrum, you can choose a much less rigorous method of breeding. If you just want to ensure that your doe keeps lactating, you can conceivably breed her to any healthy buck—of any breed of goat. If you're planning to sell the resulting kids at a livestock auction where goats are sold by weight, however, you should try to breed your does to large bucks. If you raise dairy goats, you might even think about breeding them with a Boer or a Kiko. We know some farmers who specifically breed their (dairy) Alpines to (meat) Boers in years that they're not planning to keep any kids. It's a winning situation all around: The Alpine mothers begin lactating, and the Boer genes add heft and bulk to the kids, so they fetch more at auction.

Breeding can be fun, and the more energy you want to put into the project, the more you can select for whatever traits you find appealing. We know a farmer who is otherwise very pragmatic but breeds black and white Boers because he likes the look of them. We know another who claims to have bred friendliness into her many generations of La Manchas. New breeds are developed with experimental crossings (the entire breed of La Manchas was developed through experimentation), and there's a lot to be said for heterosis, or hybrid vigor. It's also good to remember that unless you're raising dairy goats, expanding your herd, or raising kids for meat, you don't absolutely need to breed your goats. If you do choose to breed them, make sure that the experience is fun for you.

WHEN TO BREED

Once you've chosen a buck, the next decision is when to breed. European dairy breeds have a fairly well-defined breeding season, from early fall to mid-winter, but meat and fiber breeds, as well as Nubians, Pygmies, and Nigerian Dwarf goats, extend their breeding period, and in some climates are fertile year-round. Does are sexually mature by the time they reach roughly 75 percent of their mature weight, usually at eight or nine months, though some farmers recommend waiting to breed until a doe is at least a year old. All female goats will go into heat for a period of about twelve to thirty-six hours, roughly every three weeks during breeding season. It does not matter whether your does are lactating, and goats are able to conceive while in milk. You'll know that your doe is in heat because she will raise and wag her tail, called "flagging," and she will often have heavy, stringy mucous discharge

GESTATION MATH MADE EASY

The chart below will give you a rough idea of when to expect kids, based on your breeding dates. Keep in mind that normal gestation varies a little between does, and can range from 147 to 156 days. The dates below are based on a 150-day gestation.

Breeding Date	Due Date
September 15	February 12
October 15	March 14
November 15	April 14
December 15	May 14

from her vagina. Her vulva may become visibly rosy and the other goats in your herd—even other does—may mount her, mimicking the role of a buck.

Unless you own a buck or have borrowed one for the season, you'll probably use the first obvious heat cycle of the year as a guide for when to breed, rather than actually inseminating the doe during her first estrus. We mark down the date of the first flagging of the season for each of our does and plan our breeding accordingly, calculating at three week intervals to come up with an idea of their cycles. Once we know approximately when our goats will come into heat, we can choose when to breed them, and we can give our breeder friends an idea of when we'll be coming. Of course, because the window of estrus is so narrow, we can only give them a ballpark range of dates and then wait to see when our does start flagging; there's no point in trying to breed a doe who isn't clearly in heat, so you really can't schedule a breeding session until she's ready. Most breeders don't mind this uncertainty, as long as you can be flexible on the day of breeding. To accommodate everyone's work schedules, for example, we generally breed our does early in the morning.

When we're deciding which estrous cycle to breed our does in, the first thing we do is take out a calendar and do the math of gestation, and I would recommend that you do the same. Kids spend approximately 150 days in utero, so look five months into the future. Will the weather in your region be frigid or temperate enough for the new kids? You don't want your first kids to be chilled or frostbitten if you can't get to them in the first minutes after birth. Does the timing coincide with any holidays or planned vacations? You may want to be assisting a goat's birth on Easter Sunday or your wedding anniversary, but if you don't, take steps to prevent it. Can you make yourself available twenty-four hours a day for the window around birth? If you have a day job with an annual busy period—mid-April if you're an accountant, for example, or final exams if you're a student—make sure that you factor that into your breeding schedule. There's no reason to add extra stress to kidding season, especially if you haven't done it before, so take precautions in the fall to make sure that your goats will give birth at a time that's (relatively) convenient for you.

 Elizabeth Kennerly steadies a skittish La Mancha doe during a breeding session at Devonshire Dairy Goats in Archer, Florida.

The first year that Karl and I successfully bred our does, we tried to breed late to avoid having kids born during a brutally cold Maine winter. Unfortunately, we missed the optimal estrous cycle, and ended up with goat kids born two weeks after I gave birth to our younger daughter. We managed, somehow, to coordinate everyone's feedings—human and goat kids—but that year's breeding was certainly not our wisest goat management decision. Ideally, you'll have kids born at a time when the weather is reasonably mild and you can be around the barn more than usual. (In most parts of the country, spring break is the perfect time.)

If selling kids for meat is something that you're thinking about, doing some research and marketing before you breed is also a good idea. Depending upon the region you live in and the ethnic populations near your farm, you might want to

consult the Islamic calendar before you breed and work backward from Ramadan and the Eid al Adha. Goats will be in high demand in most Muslim communities during these holidays, and if you can time your breeding so that the kids are good-sized at that point, you'll have a ready market. Easter kids are also desirable in many Greek and some Italian communities. A Web search and a few phone calls in the middle of summer can help you find a market that may influence the exact heat cycle in which you choose to breed.

INSEMINATION

Once you've chosen a buck and a period of estrus, you'll be ready to prepare your does physically for pregnancy. About three weeks before you breed your does, you will start to "flush" them, meaning that you'll increase their nutrient intake. Does that are neither pregnant nor lactating do not technically need the energy boost of grain (we spoil ours and give them grain year-round, but this may be unnecessary for their nutritional needs). Extra nutrients help their bodies prepare for pregnancy, however, and increase the chance of conception, so for the last heat cycle before they are bred, slowly start giving your does extra rations of grain or high quality forage.

The buck has been chosen, the does are ready: Now you arrive at the physical act of breeding. You have several options for the actual insemination, and I'll touch on each briefly. At the most basic level, the choices are: an extended visit, during which the buck is in with the doe (or does) for an entire heat cycle; a single managed breeding session; or artificial insemination. A month-long period of cohabitation requires the least management (at least in terms of the actual breeding), while artificial insemination requires the most knowledge and skill. At our farm, Karl and I choose the middle ground, which I'll explain in the greatest detail, but first a word about artificial insemination.

ARTIFICIAL INSEMINATION

AI, as it's called, involves the insertion of semen directly through a doe's cervix and into her uterus. If done correctly, the success rate for conception is between 50 and 75 percent. The greatest advantage of artificial insemination is that it allows you to breed your doe with semen from any buck you choose, no matter where he lives in the country—and even to those that are no longer alive, but whose semen was collected before death. While a semen storage tank is a big investment, if you're interested in taking your breeding program to that next level, the expense can be worth it since it opens up your options considerably. Straws of frozen semen can be stored for months at -320° Fahrenheit, though once they're thawed, the sperm is viable for a maximum of twelve hours (as opposed to unfrozen semen, which can be viable for up to two and a half days in the doe). This means that you must be very precise about when your goat is in estrus. Artificial insemination is not for the untrained, and there are clinics and workshops around the country that offer courses in AI (see Appendix D). Some veterinarians will perform the actual insemination, and there are also services in which, for a fee, a professional inseminator will come to your farm. You may end up artificially stimulating your goat's heat cycle with hormones, or synchronizing your does (also with hormones) so that they're all in heat at once in order to perform AI. Medications of this kind can be costly, so make sure you do some cost/benefit analysis before you decide on AI. Artificial insemination can be a wonderful tool for your breeding program, but it also involves quite a bit of coordination, which should be taken into account before you make any decisions about insemination (or invest in a semen storage tank).

Breeding Sessions

Our preferred method of breeding is to give our does a quick session with a buck. We watch for flagging and begin feeding our does extra grain at the first sign of heat. Three weeks later, in the following period of estrus, we take our does to a farm an hour away to be bred, making sure that they are calm and comfortable in transit, since in goats, as in humans, stress can make it more difficult to conceive. Once we arrive at our friends' farm, we take the doe to their barn on a lead, and leave her clipped into the leash. When the buck is brought in and courtship begins, Karl keeps a hand on our goat's lead or collar to help steady her, as needed, while she's being bred. The buck blubbers and stamps and displays the flehmen response, pulling back his upper lip to sniff the doe's sex hormones with his Jacobsen's organ, located in the roof of his mouth. If the doe urinates, he may try to catch it in his mouth (something that becomes harder to explain to our daughters every year). The buck circles the doe until she's receptive, which she signals by looking back at him with an almost coy expression, then mounts her and ejaculates, sometimes hopping and twisting around as he does. When he's finished, he dismounts and stalks around, sometimes sniffing the doe's labia and licking his own genitals. After a few minutes, we let the entire dance begin again, for the insurance of a second round. When it looks as if she's settled, or con- ceived, we lead her back to the car, give her some extra grain and gentle pats, and take her home. The entire session takes about half an hour, and we've been known to drop a very fragrant Karl off at the office on our way home. Interestingly, when we bring the first doe back to the barn, smelling strongly of buck, it seems to speed up heat for the others, and their cycles sometimes shorten by a few days.

Once you've bred your doe, mark the date of breeding and the buck you used in your records, as well as the estimated due date, 150 days later. We also note the breeding and due dates on the family calendar, in the kitchen, so that we can start the countdown to birth.

PREGNANCY

For the new farmer, it can be very difficult to determine whether a goat is pregnant. There are certainly signs to look for, but to the novice, even these can be deceptive. Most small-scale farmers, ourselves included, try to determine pregnancy themselves, accepting the fact that sometimes they will be wrong. Your goat will give you pretty clear signals, but it takes a little experience to learn to read them. It's important to watch your goats closely throughout gestation and prepare for kidding, even if you're not 100 percent sure that your does have been successfully bred. The last thing you want is to begin your first experience of kidding without the necessary equipment and preparation. Though there aren't "First Response" pregnancy tests for goats on the market, if you really need to take away the mystery and know for certain that your doe is with kid, you can ask your vet for an ultrasound. Many have mobile ultrasound machines, and if yours doesn't, we always find that it brightens the vet tech's day to see us leading our goats into the office. (After clearing the waiting room of dogs, of course!)

The first sign to look for when trying to determine pregnancy is a missed estrous cycle. Watch your doe when she would next go into heat. If she's flagging again three weeks after she's been bred, she probably did not conceive and you can try again on this cycle. If she's not flagging, chances are good that she's settled.

Another way that some farmers can tell early in pregnancy is to feel their does' abdomens, right in front of their udders. The belly of a doe that has been recently bred should feel taut, while an open doe feels softer. This can be deceptive, however, as many does will automatically tense up if you touch their bellies. To be perfectly honest, Karl and I haven't mastered this technique, but we've heard from farmers who have that once you have the feel for it, it's pretty accurate. If your doe is lactating, you'll notice that her milk production drops off around the first missed estrus, another good indicator of pregnancy.

Until near the end of gestation, by about three and a half months at the earliest, the expansion of girth that marks your doe's pregnancy can easily be mistaken for the normal abdominal flux of rumination. Especially if she's carrying a single kid, a doe can mask pregnancy for almost the entire gestation. The more kids she's carrying, of course, the wider she will be, and as you get to know your goats' bodies, you'll find that each breeding season you'll be more confident of pregnancy. As pregnancy progresses, the doe's abdomen will fill out on her right side (the left side is filled by her rumen). In the later months, you should be able to feel kids if you palpate your doe's belly. Again, you will place your hands in front of her udder, feeling up the sides of her belly with your fingers. Gentle pressure of your full hand is all that is required—try not to squeeze her or make her uncomfortable. Just as in human pregnancy, you'll be able to feel the kids swimming in their amniotic fluid as they get larger, and they will occasionally kick. It's impossible not to get excited about kidding season when you feel those amazing little flutters!

At our farm, Karl and I treat every doe that has been bred as though she's pregnant, whether we have the confirmation of a kicking kid or not. We learned our first year that just because a doe isn't acting pregnant, it doesn't mean that she isn't. Our stoic Chansonetta didn't let on that she was ready to kid until she actually began to push—luckily our kidding kit was still handy. We do know some farmers who have ultrasounds done in order to make more precise management decisions—such as isolating pregnant does and increasing their grain rations without feeding more to the entire herd—but on a small scale like ours, we don't feel that the expense makes sense. Even if we feed the entire herd an extra quarter pound of grain a day for the entire pregnancy, on our scale, the cost of grain will still be less than having one ultrasound performed.

Treating goats as though they are pregnant entails the following: Once we've bred our does, we increase their grain rations to about a quarter pound at each feeding (if they're not lactating; if they're still in milk, we calculate the grain rations based on milk production). Adding to their daily nutrient intake is essential, but the amount you give is a personal choice—we know some farmers who give less, feeding pregnant does a maximum of a quarter pound per day, but we don't know many who give more. You don't want your doe to be carrying extra weight around when she kids, or to have any health problems associated with over-feeding. So while she needs some extra nutrients, make sure that you don't become too indulgent!

About two months before she's due to kid, you should dry off any lactating does. Plan for this in advance so that you can do it gradually to minimize the risk of inflammation to the udder. (See Chapter Eight on milking for specific instruc-

tions.) If she's been lactating for the entire year, your hardworking doe could use a break, and you'll want her to be dry for forty-five to sixty days before she kids.

In general, while you don't want to coddle your goats too much, you do want to give your pregnant does some extra care and love. Their bodies have gone through enormous changes during pregnancy, and they'll go through even more during labor. Now is the time to show them gratitude for what they bring to your farm (and for the milk you may be anticipating)—a little extra hay, some soft bedding, and a few more scratches behind the ears are all good ways to show affection.

If your goats have a strongly defined hierarchy within the herd, you'll notice that the dynamics of the paddock can change drastically when your does are pregnant. We keep our wethers and does together, and while our wether Joshua is still the lord of the yard, Chansonetta and Percival vie for lieutenant status when Chansonetta is pregnant or lactating. Though she is usually the more dominant goat, he senses her weakness and distraction and uses the opportunity to struggle for power. The clashes can become pretty brutal, and we've found that isolating the does is the most effective intervention.

No matter what the dynamics of the herd are, a couple of weeks before they are due to kid, you should separate your pregnant does and put them in the kidding pen to get ready. Be sure to allow supervised exercise during these last two weeks. Many heavy goats find it easier to lie around without making an effort to eat. A missed meal during this last stretch, however, could be critical. If there are struggles for power, especially if they seem likely to end in an injury that could threaten the pregnancy, you may choose to isolate them earlier. As with everything, we find it best to keep an eye on the situation and when in doubt, be cautious. An ounce of prevention, as they say. . . .

A pregnant Boer doe grazes on tall grass at Leslie Wootten and Jerry Baldwin's Dalco Farms in Casa Grande, Arizona. Leslie loves the daily chore of feeding her goats. "In this world where I feel so insecure, where you never know what's going to happen next, I love the fact that I can go out in the morning and they're always there, ready to eat. It doesn't matter what else is happening in the world, they need to eat. They want to eat, and I want to feed them."

CHAPTER SIX

Raising Kids

No matter how many times you've done it, helping a doe deliver her kids is a thrilling, moving, nerve-wracking experience. Even if everything goes smoothly, you'll probably be anxious until the last kid is out. If it's your first kidding season, the arrival of baby goats is only the beginning of an intense time, a period when you may find goats living in your kitchen, a dryer that's perpetually warming towels, a fridge stocked with bottles of colostrum, and feeding schedules tacked to every wall. This goat invasion is perfectly normal, and as you spend more time with goats, you'll begin to anticipate the chaos of kidding season as eagerly as you look forward to the arrival of the kids. In those rare moments when you can pause and catch your breath, you'll find the whole experience enormously rewarding—especially the delightful serendipity of the season. There's truly nothing like the sight of a wobbly kid prancing around your living room with the family dog!

The majority of the time, kidding will progress smoothly, with your laboring doe needing very little assistance from you. On the rare occasions that your goat needs help, you should contact your veterinarian or goat mentor immediately. While I'll discuss the warning signs of a difficult labor, the recommendations in this book should not be a substitute for experienced medical advice. While most commercially sold kidding kits include scalpels and obstetric leg snares, you

A Spanish doe cleans a pair of newborn Spanish/Boer cross kids at Larry and Peggy Krech's Kevuda Haven Ranch in Center Hill, Florida.

should not attempt to use them without having
had live instruction. Unless you have had
hands-on training, offering your goat the
wrong kind of help during labor can have
potentially serious consequences. When in
doubt, close the books and call an expert.

Godfrey, the author's dog,
watches as a two-week-old
Alpine kid gets ready to
jump from the sofa in the
farmhouse of Ten Apple
Farm in southern Maine.

    That said, in most cases, your goats will
deliver on their own, with you present solely to provide moral support and a big
bucket of warm molasses water when the last kid is finally out. Your work begins
when the kids slip onto the shavings and need cleaning, attention to the umbilical
cord, and, if you're bottle-feeding, their first sips of colostrum.

Preparing for Kidding

Goat gestation takes approximately 150 days, but as with human birth, every pregnancy is different. Some goats kid a little early, others kid a little late; diet and age
affect gestation, but it's also influenced by the goat's natural rhythms. Our does, for
example, tend to kid a day or two after their due date. A normal goat pregnancy
can last from 147 to 156 days, which leaves a window of ten days when you need to
be on call. As the due date approaches, you should start assembling your kidding
kit (see sidebar on page 84) and watching your doe for signs that labor is immi-
nent. You should also prepare the kidding pen, a stall where your laboring doe can
be by herself while she kids, and where she can recover from the exertion. If you're
planning to leave the kids on their dam, the kidding pen is also where they will
bond with their mother for the first day or two after birth. The kidding pen should
be at least five by five feet, though ours is larger so that the doe has plenty of room
to pace during labor. The bedding should be fresh and clean: Shavings are nice
because they offer a soft cushion to the new kid, but straw is also good because it
won't stick to the baby's damp skin. Either way, the pen should be clean, with a
feeder stocked with your best hay, and, as soon as the doe is in, a regularly replenished bucket of cool water.

    If you have an off-farm job, it's also a good idea to give your colleagues a
heads-up that you may be abruptly called away, and to get ahead on any work you

can. If your goats decide to kid at night, you may be sleep-deprived for a few days, and even if they kid at a reasonable hour, you'll probably be distracted.

CHANGES IN YOUR DOE

As your doe gets close to freshening, you'll begin to notice physical and behavioral changes in her. Kids put on up to three-quarters of their body weight in the final six weeks of gestation, so in the last month of pregnancy, your doe's body will widen noticeably. Her udder will begin to fill out and become engorged—sometimes stretching so much that the skin appears shiny—and her vulva may become reddened and slightly protruded.

A week or two before she actually kids, the pelvic ligaments above the doe's hind legs begin to soften. These ligaments are located over the hips, angling back and out from either side of the spine. Normally they are taut, round, and roughly a centimeter in diameter. As birth approaches, the release of the hormone relaxin

softens the ligaments until there's a visible indentation on either side of the spine.

    Every doe's experience of labor is different, but there are several common signs that kidding is about to begin. Immediately preceding birth, in the final forty-eight hours before kidding, your doe may isolate herself from the rest

Flyrod inspects her first kid, a single buckling, shortly after birth at Ten Apple Farm in southern Maine. The kidding pen should have straw or clean shavings and plenty of fresh water for the doe.

of the herd or wander off to a private corner of the paddock. She may begin to bleat softly and show no interest in her food. Her vulva, which may have been

bulging, will elongate, soften, and discharge heavy white mucus. If you haven't already, this is the time to confine your doe to the kidding pen.

In these last days leading up to kidding, you should keep a close watch on the barn. If it's possible for you to be at home, stay there and check on the barn hourly. If you have a baby monitor, put the transmitter in the pen (out of reach—even heavily pregnant does will nibble through a wire, if given the chance) and keep the receiver with you. A doe that's going into labor will bleat and moan as labor intensifies. Some farmers have a "barn cam" that allows them to watch the barn from the comfort and warmth of their home. These can be purchased relatively reasonably from many goat supply companies, and if this seems like a worthwhile investment to you, it will give you an accurate view of what's going on with your goats. A word of caution, though: Barn cams can become very addictive!

If it's not possible to stay at home, keep as close to the barn as you can, and if you have neighbors or friends who can help with the vigil, ask them to check in on the goats at regular intervals. One year we were lucky enough to have a nine-year-old friend visiting during kidding season, and it was Erica, when the rest of us had gone to bed, who made one final trip to the barn and discovered a laboring doe. By the time I was dressed, the head was already out and Erica was whispering soothing words to the goat as she gave her final push.

LABOR AND BIRTH

Pre-labor, with its gradually intensifying hormonal changes, should last anywhere from twelve to thirty-six hours. During this time, your doe will become progressively more uncomfortable, and often more vocal. She may begin digging in her bedding to make a "nest" for her babies. She may stretch and yawn, shifting her

KIDDING KIT ESSENTIALS—WHAT YOU SHOULD HAVE ON HAND

By your doe's fourth month of pregnancy, you should organize a kidding kit to have at the ready for birth. Both Caprine Supply and Hoegger Supply Company (see Appendix C) offer already assembled kidding kits, which are very helpful when you're trying to make sure you've covered all your bases. Below is a list of what you should have on hand during kidding season:

Iodine tincture—for dipping newborns' navels and the bottom of their hooves

Film canisters or shot glasses—to hold iodine tincture during navel dipping

Sharp scissors, stored or dipped in alcohol to sterilize—for trimming the end of the umbilical cord

Cotton string or unflavored dental floss—to tie off end of umbilical cord

Plastic digital thermometer—for does and kids, a healthy temperature should range between 102 and 103 degrees Fahrenheit

Infant bulb syringe—to siphon mucus from kids' nostrils or throats

Obstetric leg snare—rubber cords that loop around the kid's legs to aid in labor; included in most pre-assembled kidding kits, these should only be used with instruction from your mentor or veterinarian

Shoulder-length obstetric gloves

Lubricant—preferably water-soluble, available in large quantity from goat supply companies, but KY will also work

Clean, soft towels—four at minimum, but more is better

Molasses—buy at least a gallon for a small herd

Powdered colostrum and milk replacer—nothing beats mother's milk, but we like to keep both of these on hand during kidding season, just in case; available from goat supply companies, make sure to buy milk-based rather than soy

Rubber nipples—available from goat supply companies, these fit onto soda bottles

Empty, clean sixteen-ounce soda bottles—run empty soda bottles through the dishwasher to sterilize (they cannot be boiled)

A camera—by your third kidding season, you may not feel as urgent a need for documentation, but the first time, you'll want to capture the moment

hips to adjust for the kids' arrival. As the time comes closer, she will probably begin to pant, taking heavy, shallow breaths with an open mouth in the caprine equivalent of Lamaze.

When labor begins in earnest, your doe will begin to have visible abdominal contractions, her belly clenching at regular intervals. She may drop to her knees on her front legs, roll to her side or bend her back legs in a slight squat, and she'll grimace and call out in loud bleats. She's in pain during these intense contractions, and it can be difficult to watch her and not want to intervene. Fortunately, with goats, once a normal labor begins it progresses very quickly, usually lasting less than an hour between the beginning of hard labor and the delivery of the first kid.

The first thing to be expelled during the onset of labor is a heavy amount of stringy, bloody mucus (in humans, this is called "bloody show"). Soon after, a white sack of clear to amber-tinged fluid emerges. This membrane can either rupture

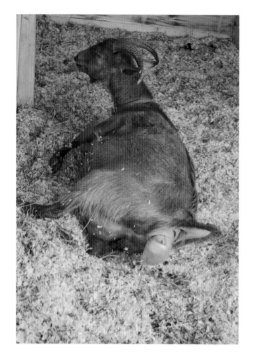

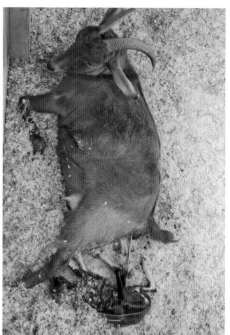

 Chansonetta, a two-year-old Boer/Alpine cross doe, gives birth at Ten Apple Farm in southern Maine. After silently beginning labor in the evening, Chansonetta was discovered in the barn as the kid emerged and she began to push.

inside the doe as the kid is being born, or can break outside the doe; either is normal. The next arrival is the kid.

There are two normal positions for delivery of a kid: front hooves first, in a "diving" position, or hind legs first. In the former, one or both of the front hooves will appear in the vulva, either within the sack of fluid or, if the membrane has ruptured, on its own. The nose of the kid will then appear between the hooves, on top of them, as though the baby is diving out of the doe's vulva. Once the head is delivered, the rest of the kid will slip out easily.

In a hind feet first delivery, the back hooves will appear upside-down in the opening of the vulva, followed by the kid's hips. While this is a normal position for delivery, in this situation the umbilical cord becomes briefly squeezed against the pelvis, cutting off its blood supply and in rare occasions tearing. If the goat appears to be having difficulty expelling the kid, it is a good idea to help the kid out with a gloved, well-lubricated hand. Gently inserting two fingers into the doe's vagina to help slide the kid out should be enough assistance. Never tug at the kid's hooves or legs, pull straight back, or exert any more than a minimum amount of pressure on the kid.

The majority of goats have twins in a normal birth, and it should take about two hours from the beginning of hard labor to the expulsion of the placenta and the delivery of the second (or third or fourth) kid. If it's been more than an hour and a half since your doe began passing heavy, bloody mucus and she hasn't delivered a kid, she may be in trouble. Other warning signs include obvious, continual pain, without a break between contractions; excessive bleeding; a breech (bottom and tail first) presentation; twins coming out at the same time; front legs with the head turned back; head with no feet presentation; or the presentation of all four legs in the vulva at once. Your vet or goat mentor (or you, if they can talk you through it) may have to help with a difficult delivery, and time is of the essence. Reproductive organs are incredibly delicate and can tear easily. If your doe is injured during birth she'll need surgical attention, so the sooner you can reach your vet, the better.

In case of an emergency, use common sense. As kidding season approaches, make sure that you alert your mentor or vet and that you have a number to contact

them at any hour. If your doe seems to be struggling, don't wait to call—it's better to wake someone up at an inconvenient hour than to jeopardize the health of your goat. Though you should not try to use them on your own, make sure that your kidding kit includes equipment such as obstetric snares and shoulder-length latex gloves, just in case. Most important, keep an eye on the barn so that you're not surprised by a laboring doe.

They're Out

The kids have been delivered, everything has gone smoothly, and you and your doe are sharing a moment of astonished elation. It's okay to give yourself about thirty seconds to revel in the birth (and, if you're like us, grab a camera to take the first family portrait), but then it's time to get to work. If there are two people in the barn, one should focus on the postpartum doe and the other should take care of the kids. If that's not possible and you're on your own, begin getting your equipment ready while the doe is still in labor so that you can work as efficiently as possible once the kids are out.

If it's been a straightforward labor, your doe will be exhausted, but otherwise won't need much help from you. She might begin nosing at the afterbirth, once it's out, and she'll probably eat the placenta—that sack of membrane that came out with each kid. This clean-up is nature's way of protecting the kids in the wild, but the placenta is rich in nutrients and protein, so even though it's not a necessary defense in the barn, if she wants to eat it, you can let her. Some farmers feel that since goats are not designed to digest meat, eating the placenta may cause digestive problems, though we haven't seen that in our barn. If she doesn't eat it, you'll need to scoop up and dispose of the afterbirth either by burning or burying it. Plan to take it out of the pen as soon as you can tell that your doe is not going to eat it; you don't want the goats to become chilled by nestling on damp bedding.

A week-old Boer kid sleeps under a heat lamp at Marvin Shurley's Shurley Ranch in Sonora, Texas. In Shurley's cement shelter with a dirt floor, the heat lamp poses little danger of fire, but one should be careful when using heat lamps in wooden barns and around hay and straw.

Your doe will need a large bucket of warm water with about a cup of molasses added to it—molasses contains iron, potassium, sodium, and calcium, and the sweetness encourages the goat to

 The author prepares an iodine dip for a newborn kid, while new mom, Flyrod, looks on at Ten Apple Farm in Gray, Maine. After tying off and cutting the umbilical cord, the kid's navel is dipped in iodine to prevent infection.

drink a lot and re-hydrate. Though she may not be hungry yet, your doe should also have her choice of your best hay. She's earned it!

The kids should be toweled off immediately, and if it's exceptionally cold, consider using a blow dryer on them to make sure they aren't damp. Though you don't want to overheat your barn, you do want to make sure that the pen in which you're keeping the kids is at least 40 degrees Fahrenheit. A heat lamp suspended out of reach can warm up just a corner of the pen, and the kids will naturally gravitate to the toasty patch.

If there seems to be a lot of fluid or mucus in a kid's nostrils and mouth, use a bulb syringe to siphon it out, and wipe its head thoroughly to help stimulate it.

Rubbing the kids' bodies and heads in gentle but vigorous strokes will also warm them up and get the blood flowing. If a kid is having difficulty breathing, pick it up by the hind legs and give it a little whack with your open palm—not too hard, just like the doctor does to new babies in the movies. This will force the kid to inhale and should kick-start regular breathing.

The umbilical cord will most likely have torn during or immediately after birth. If there's more than an inch dangling from the navel, snip it with sharp, sterilized scissors; if it bleeds, tie it off with cotton string or unflavored, unwaxed dental floss. Fill your film canister or shot glass with iodine and dip the navel, being careful to cover the area fully—we slosh the iodine around to make sure, which gives the kids bright yellow tummies, but ensures that the navel has been bathed. If you are going to put your kids outside immediately or raise them with their mothers, it's a good idea to also dip their hooves in iodine.

While you have them, you should weigh the kids, either in a kid sling on a hanging scale, or by holding the kid on a bathroom scale and subtracting your own weight. It's important to get a baseline weight so you can gauge the growth of your kids. Generally, a healthy goat kid will weigh roughly the same as a healthy human baby, in the six to nine pound range.

Deborah, Jonathan, and Catherine Boehle weigh Carmen, a two-week-old Nigerian Dwarf kid, at Antiquity Oaks in Cornell, Illinois. Carmen almost died from hypothermia because she was the last of triplets and her dam was too busy with the other two kids to get her cleaned up. Luckily, the Boehles found her soon after birth and were able to bring her back, but only after tube-feeding her for twenty-four hours. Carmen is now five and in 2009 kidded with twin does.

FEEDING THE KIDS

There are many ways to manage feeding new kids, and it's important to know at the beginning that you have options. Removing kids from their mothers to bottle-feed them is one possibility, and it's most common in dairy herds, where the farmers are interested in keeping some milk for themselves. It's also preferred because research has shown that if you pasteurize milk before you feed it back to the kids, it helps control any diseases your herd might have that are transmittable through mother's milk. This will be discussed more fully in the chapter on goat health, but for disease prevention, many farmers immediately take kids from their dams and pasteurize the colostrum and, later, the milk before bottle-feeding.

Other management options include leaving the kids on their dams exclusively, which is generally done in herds raised for meat and fiber. A third option, sometimes done in healthy, non-commercial dairy herds, is to leave the kids on their dams for part of the day, but then to separate them for a long enough stretch (six to eight hours) that the does can be milked. The advantage to this method is that it allows you to keep some milk and accustom the kids to you through bottle feeding, but it also gives you a little more freedom during the first months after kidding, when the babies need to eat four or five times a day.

If you're planning to leave the kids on their dam, make sure that they begin nursing within the first half hour after birth, preferably sooner. Some goat handbooks recommend squeezing a small amount of colostrum into a strip cup (a small

stainless steel or plastic cup with a screened cover) to make certain that the doe is lactating, even if you're not planning to milk her. If you're raising goats for meat or fiber and they will let you hand-milk them, this is a great idea, but we've heard from farmers that this can be a frustrating exercise because goats that aren't used to having their udders handled often won't "let down" their milk. If you can't hand-milk them, keep watch over the kids and their dam to make sure that she's letting them nurse. In a successful feeding session, you'll hear a faint, rhythmic gulping from the kids, and, as they get the hang of it, you'll see foamy bubbles at the corners of their mouths and a tiny milk mustache when they pull away.

If you're raising dairy goats and are planning to bottle-feed the babies, take the kids away as soon as possible to avoid imprinting. (Imprinting is the early process by which kids and their dams learn to recognize each other

A Boer doe bleats loudly after giving birth to triplets at Marvin Shurley's Shurley Ranch in Sonora, Texas. Shurley, the President of the American Meat Goat Association, has raised Boers since the early 1990s and is a strong advocate for the breed.

 The author's two-year-old daughter, Charlotte, and Eleanor Sterling, age four, bottle-feed the baby goats during Charlotte's birthday party at Ten Apple Farm in southern Maine.

by sight and smell; research suggests that if kids are taken away soon after birth, imprinting doesn't have a chance to occur.) Separating a baby animal from its mother can be heart-wrenching, but it's necessary if you're serious about milking, as otherwise the kid will continue to drink and you'll be left with less milk.

When our first kids were born, it was two weeks after I had given birth to our younger daughter, Beatrice. Though we raise dairy goats, taking the kids away was out of the question; it was simply too emotional for me. Our doe's labor had begun while Karl was at work, and my girls and I coached Flyrod (the dam) through the delivery by ourselves, hovering in the barn and shouting the goat's progress into my cell phone to Karl as he sped home. Having just done it myself, I felt very connected to the doe as she labored, even more so because my own new "kid" was nestled into my chest in the baby wrap. By the time Karl arrived in the barn, the buckling was out and our daughter Charlotte, age two at the time, had hopped off the hay bale she'd been perched on during labor and was helping me towel him off. Though Karl and I had discussed bottle-feeding the kids, when the buckling was actually born I projected my own feelings onto the goats and couldn't bring myself to take the kid away from his mother. After three hours, however, it was clear that she had no interest in the kid. Though she had sniffed him when he first came out, she blocked him with unmistakable hip-checks when he tried to nurse. Finally, I took him out of the pen, hand-milked the goat, and began bottle-feeding the kid colostrum. Flyrod never batted an eye and has been a great milker ever since. The moral of this story is that as easy as it is to project your own feelings onto your goats, it may do more harm than good. Your goats need strong, unsentimental management from you, even if that means doing things that make you uncomfortable.

If you leave the kids on their dams, once it's clear that they're nursing well you can assume that they'll keep nursing until they begin to eat hay. Fresh water and hay should be provided for them, along with the rest of the herd, and you can expect their mothers to wean them gradually so that they're eating solid food exclusively by about two and a half months. This should include approximately half a pound of grain a day by their sixth week, working up to a pound of grain a day by two months. They should continue with this supplement of grain until they're a year old. You'll want to spread the grain out between their feedings so that they don't gorge on it all at once, preferably dividing it evenly between their morning and evening feedings.

The time frame for "bottle babies" is roughly the same, but in their case, you'll be the source of their food. The chart in the sidebar (see page 97) gives an optimal schedule for bottle-feeding kids. In our house, we make a chart for each kid and tack it to the wall of the kitchen (close to the refrigerator full of milk bottles) so that we can record each feeding. We generally feed the kids first thing in the morning, around noon, during evening chores, and one last time before bed, spacing the feedings about five hours apart. Keep in mind that for the first two weeks the kids will be eating at least four times a day, in feedings spaced at regular intervals, so make sure that your schedule can accommodate at least one midday session.

When bottle-feeding a kid, you will need to warm the milk to body temperature (approximately 100 degrees Fahrenheit) for ease of digestion. This can be done by simply filling the bottles and setting them in a bowl of hot tap water for a few minutes. If you're pasteurizing the colostrum or milk beforehand, make sure

BOTTLE-FEEDING SCHEDULE

Below is a sample feeding schedule for new kids. The quantity of feed is the optimal maximum that your kids should eat. Some kids will want more, while it will be a struggle to convince others to take any. If a kid isn't drinking its full bottle, we'll often add an extra feeding to make sure that it's eating enough. At the other end of the spectrum, if you're using a feeding bucket or a shallow pan and you have a kid that's gorging itself, it's a good idea to bottle-feed all the kids once a day to make sure that everyone's getting the right amount of milk. As with everything, watch your kids and use common sense.

At our house, we make blank schedules with space for the date, the feed, the time of feeding, and the amount fed to each new kid. These schedules are tacked to the wall to record feedings for the first eight weeks. Not only do they give us a record of the kids' appetites, they also keep us from unintentionally doubling up on feedings.

A two-week-old Alpine kid peeks out of his cardboard box inside the farmhouse at Ten Apple Farm in southern Maine.

Age	Feed	Quantity	Number of feedings
Birth–2 days	colostrum	4-5 oz per feeding	4–5 feedings daily
3–7 days	milk	8-10 oz per feeding	3–4 feedings daily
2–6 weeks	milk	12 oz per feeding	3 feedings daily
	water		free choice
	hay		free choice
	grain		begin to introduce each feeding
6–12 weeks	milk	12 oz per feeding	2 feedings daily
	water		free choice
	hay		free choice
	grain	$^1/_2$ pound per total	2 feedings daily
3 months–1 year	milk	0	(wean)
	water		free choice
	hay		free choice
	grain	1 pound total	2 feedings daily

that it's cooled to body temperature before feeding time. Colostrum thickens and becomes the consistency of pudding when heated to normal pasteurization temperatures. To pasteurize it, you'll need to heat it to 135 degrees in a double boiler and hold it at that temperature for an hour. Do not heat it above 140 degrees, or you'll risk killing the beneficial antibodies that make colostrum so essential to the health of your kids. To pasteurize milk, heat it to 145 degrees and hold it at that temperature for half an hour, or heat it to 165 degrees for thirty seconds. This can be done directly on the stove (rather than in a double boiler), and the milk should be stirred regularly to prevent scorching.

Before feeding, measure your colostrum, milk, or milk replacer into a clean glass measuring cup and pour it into an empty, clean soda bottle, topping it with a rubber nipple (see feeding chart). You can either warm the fluid before filling the bottle, or warm the bottles after they're assembled (we generally do the latter, keeping a few ready-to-go bottles in the fridge at all times). At the beginning, you may need to take each kid out of the pen individually and hold it on your lap for feeding,

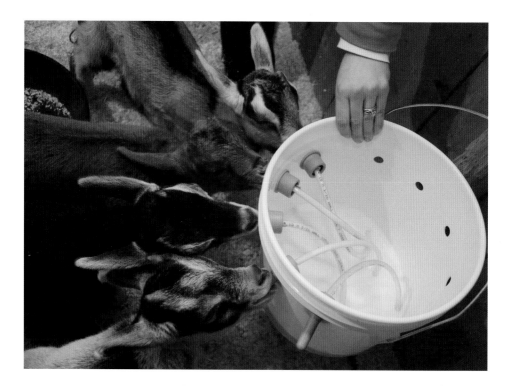

especially if it's a reluctant eater. Later, as the kids become more comfortable with the process, you can either go into the pen or just reach over the side to feed them. Alternatively, after the first week or two, you can begin

The author steadies a bucket of milk for four month-old Alpine kids at Ten Apple Farm in southern Maine. Once kids are trained to suck on rubber nipples, a feeding bucket like this can be a real timesaver, instead of bottle-feeding kids individually.

feeding the kids their milk in a clean, shallow pan, set on the floor of their pen. It may take them a little while to figure out how to drink this way, but once they've learned it will encourage them to drink from the water bucket and may simplify weaning. You can also invest in a feeding bucket: a plastic bucket with holes drilled around the top or the bottom. In the former model, each hole fits a nipple that connects to a plastic tube so that kids can suck up milk from the bottom of the bucket. In the latter, the nipples are fitted directly into the bottom of the bucket. Both types allow you to feed many kids efficiently and are available in various styles from most goat supply companies.

When feeding the kids, tip the bottle upside-down so that the goat cranes its neck back. This position mimics the way a goat nurses on its mother, and makes it easy for the kid to latch onto the rubber nipple. If the goat's having trouble sucking, you can give the bottle a gentle squeeze, being careful not to force too much into the kid's mouth. (You'll know if you've squeezed too hard because the kid will begin to sputter and spit.) You can also rub the kid's neck and under its chin to encourage sucking. We keep a few styles of nipples around, too, and try different ones on kids who are having trouble getting the hang of the bottle. One year, we were afraid that we were going to lose a kid who refused to take the bottle, and it was our little Charlotte, just about the goat's size, who got him to drink. Whatever you do, stay calm and stay at it. A little patience and good humor goes a long way!

As with human babies, while bottle-raising goat kids you'll gradually begin to offer solid food, introducing grain and hay while you cut back on the amount of milk you're giving them. Until they're six weeks old, anticipate that milk will make up most of their diet, but by about a month, make sure that they have unlimited access to hay and a regular offering of grain. Hay and roughage will start to stimulate their developing rumen, and the earlier they learn to eat it, the sooner their adult digestive system will begin working. When you begin to wean, around seven or eight weeks, or when the kids are three times their birth weight, begin replacing some of their milk with water, increasing the amount over the days until they are no longer drinking any milk.

Kid Management

In addition to feeding the kids, in those first few months you'll need to tattoo them and, if you're disbudding and castrating them, now is the time. If you're registering any animals, contact your goat association as soon as possible to make sure that you're up to date on their requirements and to get your paperwork in order.

Tattooing

In the United States, livestock is tattooed as a means of positive identification. Alternatively, many meat and fiber goats are identified with ear tags, and some

☛

The author holds Percival, a one-month-old Alpine buckling (soon to be wether) kid, and gets ready to take the goat home after picking him up at Appleton Creamery in Appleton, Maine. The green inside his ears is from tattooing.

farmers are moving toward implanted microchips in the ear or tail web. As of this writing, the tattooing or tagging is not required by the U.S. Department of Agriculture (USDA) and is strictly voluntary, but it is required of any animal that you intend to register, show, or have appraised. As the USDA begins implementation of its National Animal Identification System, some method of positive identification may become mandatory, even for the small-scale farmer. It's a good idea to tattoo or tag your goats as a matter of habit; it's a simple process and will keep you on the right side of the rules if more regulations are enacted. Inserting ear tags is relatively straightforward, but tattooing requires a bit more instruction. Goats are generally tattooed in their ears, with the exception being La Manchas, which are tattooed in the tail web. The process of tattooing involves making a shallow puncture of the skin with tattoo tongs, then smearing the puncture with ink. Tattooing ink comes in liquid or paste and is a vivid green, which will leave your kids looking like leprechauns for a few days until the excess wears off. Each registration association has a specific preference for the way goats are tattooed, but generally your herd code will go in the kid's right ear (or the right side of the tail), and its individual number will go in the left. Make sure to familiarize yourself with your registration association's rules (if you have one), or, if you don't plan to register your goats but do want to tattoo them, check with your local extension officer to see if your state has a preference.

Tattoo kits are sold by most goat supply companies, and they will include specific instructions about using the equipment. Cornell University's animal science program has an excellent slide show that demonstrates how to tattoo a kid; you can find it online at ansci.cornell.edu/4H/meatgoats/tattooing/tattoo.htm.

The most important things to remember while tattooing are:

- Sterilize all equipment before and between use. Dip everything—numbers, brushes, tongs or pliers—in an alcohol bath before and after each use.

- Tattoos are permanent. You don't want to use a B instead of an eight, or to accidentally flip a number. Test each code on a piece of paper before you actually imprint it on your goat.

- Tattooing is a two-person job. Ears are extra-sensitive parts of the body, so your goats will be briefly uncomfortable and will squirm while you're tattooing. Make sure that you have a helper to restrain the kid while you're working.

DISBUDDING

I love goat horns. They are elegant to look at and, as natural radiators, they let goats adapt to their climate by regulating their body temperature. They are also a goat's primary means of defense, which is another way of saying that horns can be very dangerous. Goats will lock horns in play, but if there are power struggles within the herd, they'll also use them on each other with a little more aggression. They will use them inadvertently, getting their heads stuck in fencing and catching each other— and sometimes you—in the barn. And they will use them with a purpose, rubbing against trees and butting against the gate of their pen if they're feeling stir-crazy.

Whether you decide to keep your goats' horns or remove them is a choice that only you can make. Some goats are naturally polled, or hornless, but the majority are born with little nubs of horn buds at the crowns of their heads, and you'll need to choose whether they are removed or allowed to grow. You should keep in mind that it's good practice not to keep horned and hornless animals in the same herd, so once you've made your decision, be prepared to stick with it. There are many factors to take into account, but the most important is your comfort level. If leaving the horns on will make you feel timid, less in control of your animals, or flat out frightened, you should consider disbudding them. You should also think about disbudding them if you're raising dairy goats and have any interest in showing them. Both 4-H and the American Dairy Goat Association require that goats have their horns removed to be eligible for showing. Most Boer associations, on the other hand, prefer horned goats in the show ring.

Horns are a part of the goat's skull, growing directly above the sinus cavities, and they have a central core of blood vessels. When horns are broken, or when an adult goat is dehorned, this core will bleed profusely. If you choose to leave the horns on your goats, know that they can break and become infected, so any injury to the horn should be treated immediately. When our goat Joshua broke off the tip of his horn on the afternoon of July 3, our vet sent us to find a tetanus booster shot at the only animal clinic that wasn't closed for Independence Day—two hours away, but longer with the holiday traffic. Our vet advised us to saw the horn so that the break was clean, so we spent the Fourth performing what felt like Civil War surgery: me holding the goat steady while Karl used a hacksaw to even off the jagged horn. Thanks to the tetanus shot and regular cleaning with an iodine solution,

Joshua was fine, and continues to be two years later. But it was a gory procedure. It's rare that an intervention like this is necessary, but if this seems like something you couldn't imagine yourself doing, horned goats are probably not for you.

If you plan to remove the horns of your goats, do it in infancy, when their horn buds are just coming in and they can be disbudded. In this procedure, a red-hot disbudding iron is applied to the head, where the horn bud is growing. The process is brief, but very painful, and most producers we know consider it their least favorite part of raising kids. Improperly done, disbudding can result in infection, brain damage, or horn skurs (horns that continue to grow at awkward angles and can cause problems later). If at all possible, I would recommend having your veterinarian disbud your goats under anesthesia. If that's not an option, take them to an experienced goat farmer who can teach you how to disbud your goats properly. Many goat guides and Internet sites give instructions on disbudding, but this is a case, like assisting a difficult birth, where live instruction is essential.

CASTRATION

Unless you're planning to keep them for breeding, or, like us, you're selling them for meat to a market that prefers them intact, when bucklings are born you'll want to castrate them relatively soon after birth. Some research indicates that urinary calculi, a potentially deadly condition in which mineral crystals block a goat's urethra and make him unable to empty his bladder, are more prevalent in wethers that have been castrated before three months of age. If you're planning to keep a wether as a pet or working animal, it's best to wait until they've passed this mark. If you're planning to sell the wethers for meat, early castration will prevent the meat from having a strong, exceptionally goaty flavor, and should be done between the ages of one and three weeks.

At whatever age you choose to castrate your bucklings, there are a few options to explore. If you're planning to keep your wethers until they reach a ripe old age and want to wait to castrate to minimize the chances of urinary calculi, surgical castration is the most effective option. It's what we did with our pet wethers, and we've been very pleased with the outcome. Like neutering a house pet, surgical castration is done by making an incision in the scrotum and removing the testicles. This is done under an anesthetic by your vet, which can be pricey, but

the procedure is painless to the goat and the discomfort of healing can be taken care of with a few days of baby aspirin.

Another method of later castration is the emasculatome, an instrument that crushes the spermatic cord, cutting off blood to the goat's testicles. If done correctly, the emasculatome doesn't break the skin, the testes eventually wither, and the scrotum remains intact but empty. Using an emasculatome, sometimes called by the instrument brand name Burdizzo, requires some skill, as the spermatic cord can move within the scrotum, and I would recommend not attempting it without instruction from a seasoned farmer.

If you're wethering your bucklings early—in their first month, as soon as their testicles have descended—the easiest method to use is an elastrator, a tool that allows you to slip a tight rubber band over the scrotum, cutting off circulation. If done without anesthetic, the kid will be in visible pain for ten to fifteen minutes (and sometimes longer), but after that the area should go numb. After banding, the entire scrotum will atrophy and fall off, leaving a flat belly after approximately two weeks. This method works best before the scrotum is fully developed. If banding, make sure to check that the band stays

Pygmy goat kids play on a log cut into a bench at River Valley Farm Stand in Oxford, New Hampshire. Kids love to climb and jump and will play for hours if there's a structure in their pen.

in place. Both emasculatomes and elastrators are available at reasonable prices from goat supply companies, but if you're only wethering a few goats per year, it might make sense to team up with another local producer and share the equipment.

There is a chance of infection with all three methods, and a tetanus shot should be administered before any are performed. The area should be kept clean and the wethers should be watched closely for signs of extreme discomfort or inflammation, either of which should be reported to your vet.

THE WHOLE HERD

There is nothing like the sight of goat kids capering about, clicking their hooves together and prancing in a lively capriole. They wiggle, they skip, and they bring joy to anyone watching them. If you've separated the kids from the rest of the herd to bottle-feed them, they will have developed their own dynamic and herd order within the group. If they've been left on their dams, the kids will make a gradual transition from their status as babies to being full-fledged members of the herd.

There are a few important things to keep in mind when integrating kids into the rest of the herd. Kids are high-energy, and you may find that they try the patience of your older goats. Be prepared to watch their elders teach them manners, nudging them with their flanks and even giving them light butts if the kids don't seem to be catching on. You generally won't need to intervene in these reprimands, but if things become too physical and the kids are getting battered, you may need to take them out and reintroduce them slowly to the herd. Also, if your goats have horns, you should remove their collars when you put them in with the rest of the herd so that they don't become tangled.

If you have unwethered males, know that a buckling is capable of impregnating a doe at two months of age. While herd dynamics make it unlikely, it is physically possible. Many producers who leave kids on their dams wean bucklings by eight weeks to avoid any possibility of unplanned pregnancy between son and mother. Babies are wonderful, but you don't want kidding season to come twice in one year!

CHAPTER SEVEN
The Health of Your Goats

When you decide to raise goats, you are agreeing to become responsible for the welfare of another living being. As independent of spirit as the animals are, if they belong to you, your goats will be dependent upon your care and judgment, and you should take their stewardship seriously. This isn't to say that your whole life needs to become goat-centric (though chances are good that it will). Rather, it means that it's your job to endeavor to keep your goats healthy, to exercise basic methods of disease prevention, and if your goats fall ill, to care for them to the best of your abilities. If you suspect that your goat is ill, you should always check in with your veterinarian or goat mentor. As a new farmer, your mind will probably always jump to the worst-case scenario, and while caution is a good thing, misdiagnosis and subsequent treatment can sometimes do more harm than good. At our farm, the rule of thumb is when in doubt, ask an expert.

A woman pets an Alpine dairy goat at the 2004 American Dairy Goat Association National Convention in Albuquerque, New Mexico. Regularly petting and touching an animal conveys affection while also letting the goat's owner do a quick assessment of its physical condition.

Goat care also includes a humane end of life—whether you're butchering them, as discussed in chapter nine, or nursing them into old age. Sometimes there are difficult decisions to make. Like parenting, animal husbandry is not always easy, but it's always rewarding.

FINDING A VETERINARIAN

No matter how much you read, and how much experience you accumulate, there will still be times when you need the help of a good veterinarian. The longer you raise goats, the less frequently you'll have to call the vet, but it's still sound practice to have backup—especially someone who has years of medical training and a license to write prescriptions. That said, it's often hard to find a vet who has training in small ruminants. If you're not lucky enough to live in "goat country," where there are a lot of farmers needing the services of someone trained on goats, or near a vet school with a goat research center, you may end up knowing almost as much about the anatomy and healthy habits of your goats as your vet.

When Karl and I first brought home our goats, we were astonished to discover that two vets in the practice we use for our dog and cat had experience with small ruminants, and one had actually lived for a time on a goat farm. This was why, for their first round of shots and our bucklings' castrations, we loaded the kids into a dog crate in the back of the car and drove them to an office in downtown Portland, nearly causing an accident each time we put them on leash and walked them

Dr. John Flood of the Brackett Street Veterinary Clinic in Portland, Maine, checks out Flyrod during a routine visit to Ten Apple Farm in southern Maine. Although his city practice rarely sees goats, Dr. Flood himself at one time raised La Mancha dairy goats.

across the street. Since they've matured, our vet comes out to the farm once a year to check on them, give booster shots, and make recommendations about their care. We're comfortable deworming and administering emergency shots on our own, but it's reassuring to have a professional opinion. We also alert the vets before kidding season, just in case we need immediate help. In a practice that's full of house pets, they get a kick out of coming to the farm, and we like knowing that our animals are unusual enough to them that they brush up on new recommendations before they treat the goats.

If a vet with experience in small ruminants doesn't happen to fall into your lap as it did ours, it may take a little more effort to find one. The best place to start is by asking members of your local goat association, your extension officer, or veteran goat farmers in your area. If these sources don't yield a good recommendation, try asking at your local farm store. They'll probably know a large animal vet in the area, and even if the practice doesn't specialize in goats, it will have experience with other types of ruminants. The Ohio State University College of Veterinary Medicine has an online directory of camelid specialists, many of whom also treat small ruminants, which can be found at vet.ohio-state.edu/351.htm/.

The bottom line is that before you bring goats into your barn, you should find and make contact with a veterinary practice that treats goats. The best way that a vet can treat your animals is to have a relationship with them when they are healthy. If your vet doesn't treat many goats, it's an especially good idea for her to see your animals in their normal state so that she can observe the distinct difference when they're ill. Don't wait until you need the vet to call one!

BASIC DISEASE PREVENTION

Healthy goats tend to stay healthy, and the best ways to prevent disease in your herd are to provide your goats with sound care, to practice general cleanliness, and to exercise common sense. Starting with healthy goats in the beginning will save you much anguish (and possible expense); your first step in preventing disease is to avoid it. In practical terms, this means buying healthy animals from herds that are free of common, chronic goat maladies. As I mentioned earlier, when you're buying your goats, don't be afraid to ask for detailed medical records and documentation of their test results. You're not being rude or overly cautious, you're just being prudent.

After buying healthy goats, make their transition to your herd as stress-free as possible—in goats, as in humans, anxiety weakens the immune system. Integrate them slowly, separating them when necessary and giving the whole herd plenty of space in which to become accustomed to each other. The middle of a New England winter, for example, when the goats spend most of their time huddled around the hay feeder and snuggling in the barn, would not be the optimal time to bring new members into a small herd. Late spring and summer, when it's warm and there's plenty for everyone to nibble in the paddock, is a much easier time for introductions.

Once you have a healthy herd of goats, basic cleanliness and maintenance will go a long way toward keeping them in good shape. A couple of simple management choices can make a world of difference: Keep the barn clean and change the goats' bedding regularly, especially in summer months. If you have portable fencing, rotate their browsing, so that they have a new spot to nibble every few days. Making both of these practices part of your routine will prevent the buildup of fecal matter and will discourage the growth of parasites. Also make sure to maintain thorough records of all treatments, vaccines, and dewormers. This is good management practice in general, but especially important if you are drinking your goat's milk or eating its meat.

HOOF TRIMMING

In the wild, goats wear down their hooves scampering over rocks, gravel, and other abrasive surfaces. On your farm, even if there are rocky outcroppings for them to clamber on, your goats' hooves will probably not get the natural filing that they need. To keep them healthy, regular hoof trimming is an essential part of goat

maintenance. Without trimming, the hard outer layer of keratin (the same protein that makes up human finger-nails) will grow long, folding over the soft frog (the hoof's sole) underneath

The author's three-month-old Alpine buckling, Percival, investigates the exam room at the Brackett Street Veterinary Clinic in Portland, Maine.

and forming pockets that trap damp manure and dirt. Untrimmed hooves will curl under themselves and can throw off a goat's gait and eventually damage its legs; the prolonged exposure to moisture can cause hoof rot, an infection that causes severe lameness. Hoof rot is treated by medication and extensive trimming to remove the infected tissue, but it's easily preventable with careful hoof care. While the combination of sharp trimmers and a large, squirming animal may seem daunting at first, the more frequently you trim your goats' hooves, the more comfortable you'll both become.

Janice Spaulding demonstrates hoof trimming during "Goat School" at Stony Knolls Farm in Saint Albans, Maine. Each year the Spauldings open up their farm and share their years of experience with Boer, Angora, and dairy goats with twenty to thirty goat owners and prospective goat owners. Programs like this, whether at a private farm or through an extension office, offer great opportunities to learn about good goat husbandry.

For the seasoned farmer, this bit of goat care can be a one-person job. In fact, we've actually witnessed an untethered, rutting buck get his hooves trimmed between breeding sessions! We're not quite so nimble, however, and at our place we trim hooves as a team. When you're getting ready, the most important things to have on hand are sharp trimmers, a jar of alcohol for sterilizing them, and a good source of light. You can also use a specially designed hoof plane or rasp to further smooth the hooves. Trimmers designed specifically for goat and sheep care are available through goat supply companies, but we know folks who use six-inch pruning shears with straight blades. The most important thing is that the trim-mers are sharp, easy for you to handle, and have blades about the length of a hoof (anything longer can become unwieldy).

We like to trim outside under natural light, so usually one of us will hold the goat while the other snips, but you can also do it in the barn by clipping your goat into a milk stand or stanchion to keep it steady. We find that it's easiest to distract the goat by trimming its hooves while the herd would normally be feeding, and we'll often do the wethers during their breakfast. You can also give treats or a little extra grain to keep them occupied. Either way, while the goat is still and restrained, pick up its hoof, bending its leg at the knee so that you can examine the underside. The ideal goat hoof looks like that of a newborn kid: squared and smooth, with the outer keratin sheath even to the frog in the center. To trim an adult, you'll hold your trimmers parallel to the bottom of the hoof, snipping back the keratin a few centimeters at a time. If you haven't trimmed in a while, there may be parts that are bent over and gnarled. If the hooves are particularly dirty or extremely dry from their environment, trimming will be easier if you wash the hooves or soak them in water (if you can get your goat to stand still). Don't be afraid to trim a lot, but stop when the hoof is a translucent white or you'll risk cut-ting into the quick, which will bleed. If this happens, which at some point it prob-ably will, spray the cut with antiseptic and keep going. Taking the pointy ends of your shears, scrape out any gunk in the frog (word of warning: it will smell bad). If there is any hoof rot present, it will look and feel like cork and should be cut back until you reach the healthy part of the hoof. If you have a hoof plane or rasp, you can use it gently to file down the keratin and the frog so that they're even. Make sure that the toe and the sole are level so that the foot doesn't hit the ground at a slant. Finally, once you've trimmed one hoof, make sure that you trim the other three. Unevenly trimmed hooves will cause your goat to walk awkwardly, and pos-sibly even to limp.

Hoof trimming is a relatively simple process and, especially if you clean the shears between hooves, it poses little risk to your goat. At our farm, however, the majority of our human injuries have occurred during hoof trimming sessions. We may be especially clumsy, but Karl and I have each had bloody run-ins with the trimmers. Be careful.

VACCINATIONS

There are a number of vaccines available for goats, and because of environmental differences, recommendations from veterinarians and goat veterans will vary by region. The two vaccinations that almost everyone agrees on, however, are tetanus and enterotoxemia, which are often combined into one vaccine called CD/T. Tetanus is caused by the bacterium *Clostridium tetani*, which is often present in fecal matter and will flourish in deep or open wounds. Symptoms of tetanus include gradually worsening stiffness, a rigid "rocking horse" pose, drooling and difficulty opening the mouth (lock jaw), and seizures. Enterotoxemia, which is sometimes called overeating disease or "pulpy kidney," is caused by *Clostridium perfringens*, a bacterium that's also found in fecal matter and soil, and affects the goat's intestines. While a balance of some of this bacteria is harmless and is actually one of the normal microorganisms found in a healthy goat's intestines, it can quickly grow out of control if a goat gorges itself or has a sudden change in its feed or the amount it eats. Symptoms of enterotoxemia include severe bloat, seizures, and foam at the mouth. Once they are symptomatic, tetanus and enterotoxemia are generally fatal. The good news is that they're preventable by early vaccination and regular booster shots.

There is currently no licensed rabies vaccine for goats, but if that's a concern in your area, some veterinarians will administer it "off-label." Make sure to include your vet early in the conversation when deciding on which vaccines are appropriate for your goats.

When you're new to goats, you may be reluctant to vaccinate them yourself, but administering an injection is something that every farmer should learn to do. Nothing beats hands-on training, and most vets will be happy to give you a lesson, but in a nutshell, this is how you give a CD/T vaccine: Fit your syringe with a twenty-gauge, one-inch-long needle and ready the vaccine, making sure that there is no air in the liquid as you draw it out of the vial. You can hold the syringe with the needle pointing up and tap it a few times to get rid of any bubbles, pushing the plunger gently until a bead of liquid swells on the tip of the needle. Once it's ready, cap the syringe to keep it sterile until you give the shot. Either clip the goat into a stanchion or have a partner hold it steady. While the goat's stationary, pinch the

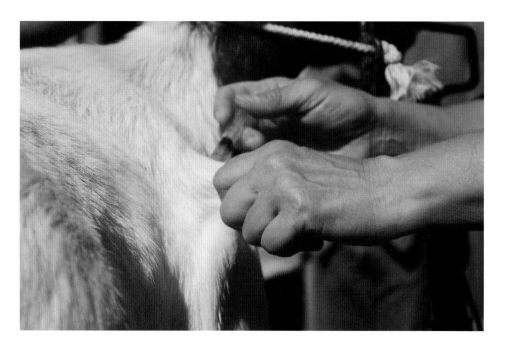

skin behind its shoulder to make
a tent, uncap the syringe and
then insert the needle so that its
tip goes between the layers of
skin (and doesn't poke out the
other side—you'll know it's time
to try again if the goat's hair
becomes wet). You have the greatest amount of control if you hold the syringe

Janice Spaulding demonstrates how to give
an injection. The proper technique involves
grabbing the loose skin around the
shoulder, forming a "tent," and injecting
the needle under the skin, making sure not
to go all the way through.

firmly in its middle. Firmly press down the plunger on the syringe with your
thumb until you've administered the vaccine, keeping the skin pinched with your
other hand. Withdraw the needle, bend it so that it can't be accidentally reused,
discard it, and you're done. Never reuse a needle, even if it's on the same goat; the
tip will become blunt, and you'll risk introducing bacteria into the animal if the
needle you're using isn't sterile.

The first time our vet showed us how to give injections, I was practicing on
our gentle giant Percival and he flinched. When I withdrew the needle, it was

bent, almost to a ninety degree angle. The goat was fine, but I was mortified. The more shots I've given, the easier it's become—I just wish Dr. Flood could see me in action now!

DEWORMING AND PARASITE CONTROL

Internal parasites are both a common and a serious hazard to the health of your goats. A small balance of parasites is normal, and actually helps goats keep up a natural resistance to them, but when an infestation develops, the results can be deadly. The larvae of mature parasites that live in the goat's digestive system are present in the animal's feces, and worms spread when browsing goats ingest these larvae. Parasites are of greatest concern where goats are in high concentration: the more poop, the more larvae. Among the parasites that attack goats are bladder worms, coccidia, hookworms, liver flukes, lung worms, and tapeworms—any of which can do serious damage and cause anemia, wasting, diarrhea, and spontaneous abortion in pregnant does. Many farmers, especially those in parts of the country that don't get a hard freeze in winter, deworm regularly as a matter of course, but as with any prophylactic treatment, there are drawbacks to exposing healthy animals to unnecessary medication. A blanket policy of medication may not target the parasites that are present, and it can cause a gradual resistance in those that are. That said, for the large-scale farmer, it may make sense to dose goats preventatively with a deworming agent, despite the risks, to maintain the health of the entire herd. For the small-scale farmer, however, there are a few more options to explore.

The most basic method of parasite control that you can practice is to maintain clean living conditions for your goats. A clean barn, fresh water, and plenty of space to browse will minimize the number of worm larvae your goats ingest. Regular inspection of your goats—checking their coats' luster, their mucous membranes, and especially their eyes—will give you an idea of their overall health. A method developed by South African veterinarians called the FAMACHA system teaches farmers how to detect anemia by matching a chart to the color of their goats' lower inner eyelids. (Contact your local extension officer to find out more about regional training in FAMACHA evaluation.) With less precision, you can watch the whites of your goats' eyes for excessive whiteness; they should actually be slightly pink, and if they're too pale, it's a sign of anemia.

The best method of internal parasite control, however, is to take a fecal sample twice yearly (generally at the end of the fall browsing season and at the beginning of spring, immediately following kidding season), have it analyzed for worm content, and work with your veterinarian to determine a deworming strategy that's best for your animals. This will eliminate unnecessary medication, and will allow you to treat only for the parasites that are present in your area. Once a regimen has been established, dewormers are generally administered orally. If you're in a hot, humid climate, you may still need to worm regularly, but if you live somewhere less hospitable to parasites, you might be able to cut back.

As far as external parasites go, lice and mites can cause discomfort, scaly dry skin, and mange, but to an otherwise healthy animal they don't pose a serious threat. If you notice your goats scratching or rubbing themselves raw, your veterinarian should be able to recommend treatment, generally in the form of a powder or cream.

COMMON DISEASES OF THE GOAT

Goats are susceptible to a number of serious diseases, and what follows is only a partial listing. This is just for reference, and if you suspect that any of these is in your herd, seek immediate veterinary attention. Don't let this group scare you—if you start with a healthy herd, exercise careful management, and minimize direct contact with other goats, chances are good that the small-scale farmer won't encounter any of these.

Bloat

Caused by overeating grain or juicy spring grass, bloat occurs when the rumen fills with gas and expands beyond its capacity. Symptoms include a stretched, inflated abdomen and visible signs of pain such as screaming, moaning, and grunting. Bloat is easily prevented by keeping grain in a secure container (we use a lidded galvanized metal trash can). In the early spring, feed grass hay to goats before putting them out on new pasture so they aren't tempted to binge.

Caprine Arthritis Encephalitis (CAE)

Often referred to as the caprine equivalent to HIV, CAE is caused by a retrovirus and, once present in a goat, is not curable. CAE can lie dormant for years and doesn't present symptoms until animals are mature. Once symptoms appear, they include swollen joints, difficulty standing, and severe wasting. Evidence suggests that CAE is transmitted through direct mucous membrane contact and mother's milk, so it can be managed within a herd. While the goats with CAE cannot be cured, the disease can be contained by isolating infected goats, immediately removing all kids from their dams, and bottle-feeding kids with pasteurized colostrum and milk. A blood test exists for CAE, and to keep your herd free of the disease, before buying a goat, it's a good idea to ask for documentation of negative CAE test results.

Caseous Lymphadenitis (CL)

Characterized by a round abscess filled with white, cheese-like pus, CL is a very contagious, incurable bacterial infection. It is transmittable to both goats and humans through contaminated pus, so when dealing with an infected goat, gloves should be worn at all times and any clothing that comes into contact with the abscess should be sterilized or burnt. Treat goats for CL by removing the abscess and quarantining infected goats until

 Deborah Boehle holds Star, a five-year-old Nigerian Dwarf doe, while veterinarian Tim Anderson prepares to draw blood. Anderson needed a blood sample to check the doe's hormone levels because she hadn't been pregnant in a year. Star is now retired but kidded last year with triplet does.

they are healed. However, they remain at risk for future infections. There is a test for CL, and as with CAE, when buying a goat, ask for documentation of negative results.

Johne's Disease

Most prevalent in cattle herds, where it's currently considered to be an epidemic, Johne's Disease is highly contagious and causes chronic wasting and eventually death. Caused by the presence of *Mycobacterium paratuberculosis*, a bacterium that can lie dormant in soil for more than a year, the symptoms of Johne's Disease are severe diarrhea and weight loss. There is no treatment for the disease, but it can be contained within herds with the same methods as CAE: isolating infected goats, immediately removing all kids from their dams, and bottle-feeding kids with pasteurized colostrum and milk.

Scours

Also known as diarrhea, scours can have many causes, from the relatively benign to the lethal. Characterized by runny, unformed feces (as opposed to healthy, dry "pellets"), which can worsen to contain mucus and blood, scours are generally a symptom of something else. Scours can initially be treated by administering an electrolyte solution, either commercially prepared or homemade. A common formula is a mixture of two teaspoons table salt, one teaspoon baking soda, two tablespoons sugar, honey, or white corn syrup, and one quart of warm water. If the goat does not respond quickly to the electrolyte solution, call your vet for a diagnosis.

Scrapie

A deadly, degenerative condition of the nervous system, scrapie is not completely understood by the veterinary community. It is considered a transmissible spongiform encephalopathy (TSE), similar to BSE or "mad cow" disease, and is the subject of an eradication program monitored by the USDA's Animal and Plant Health Inspection Service (APHIS). Many goats and sheep are ear-tagged for identification by this eradication program, and herds that test positive for scrapie are destroyed. The disease is extremely rare in goats—the most recent public report by APHIS records only twelve documented cases—but because the cause is unknown, there are no recommended tactics for prevention.

BIOSECURITY

Discussion of disease brings up the issue of biosecurity, which can be a pretty hot topic. In general, it's a good idea to develop some of your own rules for biosecurity on your farm. While you don't want to slip into the realm of paranoia, you also don't want to ignore very real threats that can affect your farm. The greatest of

these is the unintentional intro-duction of pathogens to your barn and soil. For the small-scale farmer, common, sensible poli-

 Tetanus and rabies vaccines readied for injection during a routine vet visit at Ten Apple Farm in southern Maine.

cies for biosecurity include providing a shallow bath of disinfectant and asking visitors to dip their shoes before coming into your barn and pastures, minimizing your herd's contact with goats from other farms, and quarantining new goats before integrating them into the herd.

First Aid

When our goats were just weeks old and had recently been given their first CD/T vaccines by our vet, three of them developed abscesses at the site of the injection. It was a fluke of the vaccine, and the abscesses weren't contagious, but Karl and I still had to lance, drain, and clean the sores twice a day until they healed. Between trimming the surrounding hair, re-opening the wounds, and cleaning out the infection, we used the majority of tools in our first aid kit. The purchases that had seemed so alarmist before the goats arrived were, we real-ized, necessities.

 Sparkle, an eight-month-old Boer doe,
lounges in the sun at Stony Knolls Farm,
St. Albans, Maine.

As thoughtfully as you may tend your goats, inevitably, at some point, one of them will fall ill. It may be an accident that leads to a broken leg, a midnight pillaging of the barn that results in bloat, or a sudden case of scours. Its solution may be something you can talk through with your goat mentor, or you may just have to keep things under control while you wait for the vet. Whatever the case, you should have a well-stocked emergency kit so that you can administer first aid. Preassembled kits are available from goat supply companies, and I would recommend purchasing one before you bring your goats home. Most will include basics such as a digital thermometer, antiseptic spray, iodine tincture, surgical gloves, and gauze and tape for wrapping an injury. Some also include syringes, needles, plastic feeding tubes, surgical scissors, and a scalpel. Make sure you have a flashlight, a pen and notepad, collars and clips to immobilize the animal, and plastic bags to dispose of any waste. Depending upon what you're doing with your goats, as you get more experienced, you may want to add mastitis treatment, antibiotics, remedy for snakebites, or other medication to your kit. All of these items can be purchased individually, but when you're new to goats, buying an assembled kit ensures that you don't forget something essential.

Expect the Unexpected

Between beginning to write this chapter and finishing it, Karl and I have scheduled our vet's annual visit, changed the bedding in both pens, mended the crack in the barn door to eliminate drafts, and continued our regular monitoring of the goats' milk for signs of mastitis. Yet this morning, when Karl put Flyrod on the milk stand and began to strip her teats, one udder was full of blood clots. With no warning and no signs of injury, we suspected mastitis but couldn't locate the cause. She didn't seem in discomfort and was her normal, placid self as we put warm compresses on the udder and slowly milked her out. The barn thermometer read 22 degrees, so we blasted a blow dryer at her legs, which were damp from the warm water we used to bathe her teat. This was the only time she seemed unhappy. We called the vet and now wait for a diagnosis, searching the Internet for clues in the meantime.

You can do everything by the book, but goats still get sick, sometimes seriously. This is a hazard of any kind of animal care, and while you know it on some level, it's still startling when it happens. When imagining your life with goats, keep in mind that there are things you cannot anticipate. Plan for the worst, hope for the best, and remember that each trial is an opportunity to learn.

CHAPTER EIGHT

Goat Milk?

After months of waiting, you've helped your does kid. Now, as they begin to lactate, you watch their udders engorge as the hay and grain you feed them is transformed into milk. They munch their oats and ruminate their cud and, hours later, each doe is producing up to a gallon a day of sweet, warm milk. After the kids have their fill, you are the beneficiary of this conversion. You realize with amazement that the goats you have been tending so diligently are suddenly sustaining you.

Milking your goats can be an intensely satisfying, even profound experience. Leaning your forehead against your doe's warm belly, feeling her udder soften under your fingers, watching your milk pail fill with rich, white foam: These quiet brackets to the day are the reason many of us got into goats. Milk for your family and the possibility of experimenting with homemade cheeses and yogurt are added bonuses.

In another, less romantic view, the structure that milking brings to the rhythm of your day is undeniable. Milking your goats forces you out of bed early, into the barn no matter the weather, and under the trance-like spell of milking whether you feel like it or not. On good days, this can be blissful and meditative; on others, it's pure tedium. Even if you love to milk your goats, there will be times when chores are, well, a chore.

The milk is weighed after the morning milking at Ten Apple Farm in southern Maine. Milk production is typically gauged and tracked in pounds as opposed to gallons.

Bill Hanks uses a small vacuum-powered milking machine to milk Tammy Wynette, a four-year-old La Mancha doe at the Matrix Compound in Gallup, New Mexico. "The small milker allows you to milk your doe quicker (in most cases) than by hand and still gives you a close one-on-one connection," says Bill.

Whether you have four does or forty, the essentials of milking remain the same. You must milk your goats twice a day in a (relatively) clean environment. For home use of your milk, this doesn't mean that the milking parlor needs to be completely sterile, or that you can't have the normal shelves full of accumulated clutter in your barn. It does mean that the floors should be swept regularly, the milk stand should be wiped down between milkings, and anything that touches your goats' udders or their milk should be sanitized before use. If you want to legally sell your milk, you must be certified by the appropriate regulatory agency in your state, often the department of agriculture. If you want to legally sell your milk or other dairy products across state lines, you must conform to the Pasteurized Milk Ordinance, the federal rule that guides milk sales. Before you begin the process of certification, you should talk with your local dairy inspector to make sure that your setup (and your idea of cleanliness) will pass muster.

Other essentials: Milking can be done by hand or by machine, and there are pros and cons to each method. While your does are lactating, you need to feed them a diet high in calories and nutrients, so you must supplement their hay and browse with grain. They must also have ready access to fresh, clean water to replace the fluids that are lost in milk production. Lastly, you should anticipate the abundance of milk that will flood your home and refrigerator and have a plan in place for its storage and use. One exceptional milker can produce more than a gallon of milk a day, and though every small farmer has done it on occasion, it's a shame to waste even a drop of it.

How to Milk—By Hand or Machine?

Everyone who owns dairy goats should know how to "milk out" their does by hand, but this doesn't mean that you need to hand-milk every day. There are advantages and disadvantages to both hand and machine milking, and while we milk our does by hand, there are certainly days when I wish we had a machine. The greatest advantage of hand milking is the wonderful connection you feel with your animals while they're on the milk stand. Hand milking is quiet, and from your stool, you can murmur soft words of encouragement to your doe as you gently squeeze her teats. Hand milking requires a bare minimum of equipment: a soft, clean rag to wipe the goat's udders, a strip cup to catch the first squirts, a pail to collect the milk, and a disinfectant teat dip to finish the session. There is nothing to break, no possibility that your morning will be thrown into a tailspin by a malfunctioning pump. Your hands-on approach means that there's little chance you'll miss signs of infection or skin irritation, so you can catch them in their earliest stages. There's also some evidence that when done properly, with hands that are thoroughly washed before milking (and between goats, if you're milking more than one), the chance of udder infection is actually reduced by hand milking.

For a small herd, the most common method of milking by machine is a portable vacuum pump. These can be electric or gas powered, and at this writing, sell new for roughly $1,000. The pump is just part of the milking system, however, which also includes a bucket with a lid to collect the milk, a pulsator to set the rhythm for milking, inflations and shells to fit over the goat's teats, and a system of tubes that draws the milk from the goat into the bucket. All told, the complete setup for a basic milking system will run, new, between $1,500 and $2,000. Used, they can be much cheaper, but as with anything, you should thoroughly check the equipment before purchase.

Milking systems can be set up to accommodate one goat at a time, or, in a commercial herd, as many as forty-eight. The process is noisier than hand milking, but it's still personal: The first squeeze or two of milk from each teat will still go into the strip cup, which you'll do by hand. Once the goat's teats have been stripped, you'll fit each teat into the milking inflation and set the pace of pulsation on the machine. When the goat is milked out, you'll dip her teats into the final cup of disinfectant. During the actual milking, while the cups are inflating with their rhythmic suction, you can scratch and rub your doe and show her all the gratitude you feel for her milk. If your hands cramp easily, or if you have arthritis or back problems, a milking machine will spare your fingers from the regular squeezing and your back from the awkward spine-torque of milking. Once your milking system is assembled, it's simple to teach others how to use it, so if you plan to travel while your goats are lactating, using a milking machine may make it easier to find a "goat sitter."

One drawback to milking by machine is the cleaning. While the actual milking is relatively quick, every part of the machine to touch either the goat or the milk needs to be sanitized before use, which can add dramatically to the time you spend doing chores. If you're milking a lot of goats, you may ultimately spend less time in the barn if you milk by machine, but if you're only milking a few, you might just end up shifting your time from milking to cleaning.

Whether you generally milk by hand or by machine, you should know how to hand-milk your goats, not only to make an informed choice, but also in case of a power outage or if your machine ever breaks. Since every machine is a little different in its design, and should come with specific instructions, I'll focus on hand milking in this chapter.

HAND MILKING

Karl and I had only milked a handful of goats when our own does freshened and began lactating. We were very inexperienced and though we'd read a lot, it really says more about the fortitude of our goats than about our own skill that we all—humans and goats alike—figured out how to make milking time successful.

About two months before our does were due, Karl found plans for a milk stand and built one along the side of the kidding pen. The platform has both a stanchion (which they can wiggle out of, horns and all) and a clip for their collars, which is more effective for getting them to remain still. We decided that we would feed them their daily rations of grain during milking, which we continue to do, so for the final months of their pregnancies, we led our does onto the stand for their morning and evening feedings. Because of their heft at the time, we added a cinderblock step, which remains, since they became accustomed to it. While the does were on the stand—both first fresheners that knew as little about their coming lactation as we did—we stroked their heads and tried to touch their udders authoritatively to get them used to such an intimate position. I'm convinced that the goats were skeptical and knew we were bluffing a little, but the repetition at feeding time did make scrambling onto the milk stand a habit.

When they actually gave birth and we began hand-milking the goats, some things went smoothly, while others did not. They hopped right onto the milk stand and allowed us to clip them in, but it took us a very long time to milk them out. Until we got the hang of milking, we often had to stop and stretch our curled fingers mid-milking, which also stretched the amount of time our poor does had to wait on the stand. Sometimes they were patient; other times we ended up with a hoof in the milk bucket and a stamping, irritated, half-milked doe. After muddling through for a couple of weeks, we hit our stride, and now we both look forward to that quiet time we have alone with the animals in the barn. At the beginning it may seem improbable, but milking time will eventually come together for even the least confident farmer.

Of course, milking has its inevitable frustrations—even now, the sound of a siren on the road or the bite of a horsefly can set a stray hoof dancing and send the milk pail into my lap—but if you're going to do it at all, it should be enjoyable. Time with your goats can be a soothing balm to the jangle of your day, and with a little advance preparation, it will be.

Hand milking requires very little equipment, but to keep your does' udders healthy and your milk drinkable, you should clean and assemble all of your equipment before you head to the barn. In general, the milking equipment should be stored together, in a clean and dust-free environment. A cabinet with a door is the best option. If that's not possible, inspect everything before milking and be prepared to rinse or entirely re-clean anything that looks compromised. Your milking equipment will include:

- a bucket of warm water mixed with either commercially prepared or home-made udder wash (see appendices)

- one clean towel per goat; some farmers use paper towels or commercially prepared udder wipes but we prefer soft, old tea towels

- a strip cup—a metal cup with a handle and a screened top into which you'll milk the first squirts from each udder

- a seamless, stainless steel milk bucket, available at most farm stores and online

- a small cup filled with commercially prepared disinfectant teat dip; we use one old film canister per goat, labeled with their initial

- hand sanitizer to disinfect your hands between does

- latex gloves, one pair for each goat you'll milk (optional)

- a hanging scale to weigh the milk, available at most farm stores; weighing the milk lets you know quickly how much each doe is producing and whether her production fluctuates or is constant

 The author milks Flyrod while Charlotte looks on. By two and a half, Charlotte was milking her first goat—squeezing the teat and getting a stream of milk into the bucket.

Your goats will have created their own pecking order within the herd, and you should take this into account when deciding the order in which to milk them. Once you've established an order, the goats will anticipate it, and after a couple of weeks, they'll queue accordingly. It's easiest to begin with the herd queen unless you have a compelling reason to shift the order. The most common reason to change the milking order is if you suspect that a doe has inflammation or infection in her udder. In that case, you should milk her last to ensure that germs don't spread.

We find that feeding them during milking keeps our does distracted and occupied, and we calculate the amount of feed that each goat gets based on her average milk production. The rule of thumb is to give a goat half a pound of grain for each pound of milk she gives. Before each milking, we fill the grain pan on the stand with the amount of grain that's appropriate for Flyrod, who we always milk first. We open the gate to the pen and Flyrod comes bounding out and onto the

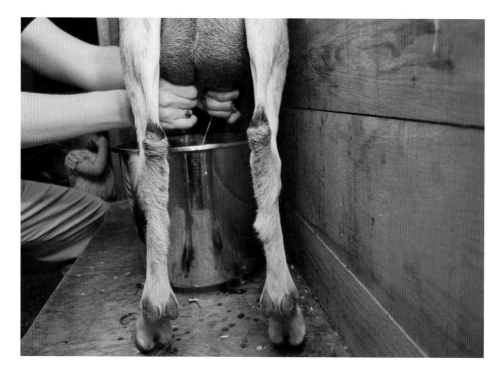

 The author hand-milks Flyrod.

stand, at which point we clip her collar into the post.

Because the orifices of their teats open up during milking, it's essential to keep your does' udders as free from bacteria as possible. To keep the udders from accumulating barn dirt, it's a good idea to clip any shaggy hair; a cordless, compact grooming trimmer is ideal for this purpose, and can be purchased at most pet stores. Once she's on the stand, we thoroughly wash Flyrod's udder with warm, soapy water. We bring a stack of cloths into the barn with us so that we can use a different rag for each goat, so as not to spread any germs within the herd. We squirt the first milk from each of her two teats through the screen of the strip cup, checking it for any lumps, streaks of blood, or a stringy consistency. Any of these signs could be symptoms of mastitis, a serious infection of the udder that makes the milk undrinkable and must be dealt with immediately. The first few squirts of milk also contain any bacteria that has collected at the orifice of the goat's teat, so it should be discarded.

Once the teats have been stripped, we position the milk pail in front of Flyrod's back hooves, directly below her teats. In human lactation they say to "bring the baby to the breast," and with goats it's the same; the teat may be longer, and it may be tempting to keep the bucket as far as possible from the goat's hooves, but stretching her teats too much can damage their tissue. When the milk pail is in place, it's time to begin milking.

To milk a goat, you make a tight ring around the base of her teat, where it attaches to the larger, sack-like part of the udder, by curling your index finger back toward your thumb. This will constrict the teat and stop the flow of milk from the udder. Gently squeeze the teat with your remaining fingers, one after the other as though you're playing scales on a piano, pushing the milk down. A jet of milk should squirt out of the teat. Relax your fingers at the top of the teat to allow it to fill up again with milk, then repeat the process until the udder no longer feels engorged and is even a little wrinkly.

If you're gentle with your doe and don't apply too much pressure to her teats, you should gradually get the feel for milking without causing her much discomfort. There are, however, a few things to avoid while you're learning:

• Never tug or pull down on a goat's teats. Just as you can cause damage by stretching them too far, you can permanently damage the tissue if you pull too vigorously. A gentle but firm squeeze should suffice to get the milk flowing.

• Don't grab too high on the teat. Teats are sacks that fill with milk, much like a water balloon. The rest of the udder contains the mammary glands, one on each side. You should be able to feel where the teat ends and the tissue of the mammary gland begins, and avoid squeezing the firmer gland. Too much pressure can cause bruising and damage to the entire mammary system. If you have trouble discerning where the top of the teat is, try feeling again when your doe is milked out; in a flaccid udder, the distinction should be clear.

• Trim your nails. If your hands are directly touching your does' udders, make sure that your fingernails are short and, of course, clean. Some farmers use gloves when milking their goats. It's a wonderful means of disease prevention, but we've found that it chafes our does' delicate skin.

• Avoid squeezing milk back into the udder. If the circle you're making with your index finger is too loose, you can actually feel milk whooshing back up to the udder. While this is bound to happen a few times at the beginning, it's uncomfortable for the doe and you should avoid it as much as you can.

You can tell that your doe is milked out when her udder is soft and flaccid. This can take as little as five minutes or as long as twenty, depending upon your proficiency. When you're close to the end, she may have a little extra milk that she's not letting down, so you can give her udder a gentle upward bump, mimicking the feel of a nursing kid, to encourage her to relax. Try to leave a generous squirt in each teat to avoid putting too much pressure on the tissue. Now is the time for some petting and affection, and I always try to rub our does' flanks and scratch their backs while they're on the milk stand.

After moving your milk pail out of the way—our scale hangs from a rafter near the milk stand, so we can weigh the pail between milkings—you'll immerse your doe's teats in a cup of disinfecting teat dip. We use a commercially prepared iodine dip, though some farmers find that this cracks and dries out their goats'

Goat soaps on display at Stony Knolls Farm, Saint Albans, Maine

 Lora Lea Misterly gives the creamy foam to her barn cats after straining the milk during morning chores at Quillisascut Farm in Rice, Washington.

udders. If your doe does have a chapped udder, you'll apply cream or balm to it after you've milked and dipped her teats. Soothing ointments and udder balms are available from goat supply companies and most farm stores; in a pinch, you can usually find udder cream in small quantities (marketed for human use) at your local drugstore.

If you're milking more than one doe, you should use a hand sanitizer on yourself (or change gloves, if you prefer to wear them) to make sure that you don't spread germs from one goat to another. You should also get a new cloth to wipe each udder and a fresh cup of teat dip for every doe.

All the fluid that has come out as milk will need to be replenished, so make certain that every doe has access to a full bucket of clean water after milking. Once they're back in their pen, our does head straight to the water bucket, and it usually needs to be filled again soon after. Hydration is key to keeping your goats healthy, especially while they're in milk.

PROCESSING AND STORAGE OF MILK

You have a barn full of relieved does, mildly aching hands, and a full bucket of milk. Now what? Goat's milk is the most widely consumed milk on the planet, and, especially when it's fresh, it's extraordinarily sweet and smooth. It is naturally homogenized, so the milk is rich and only develops a light layer of cream after a few days in the refrigerator. Its fats and proteins are slightly different than those of cow's milk, and some people who have difficulty digesting cow's milk have no trouble with goat's. The many vitamins and minerals in goat's milk are also absorbed through the skin from goat's milk lotions, and naturally occurring caprylic and alpha-hydroxy acids make goat's milk soap a gentle cleanser. In short, it's wonderful stuff.

When you're fantasizing about your life with goats, an abundance of milk seems a glorious thing. But when jars of milk have taken over your refrigerator and you're wedging leftovers into the crisper, you may think otherwise.

Before your goats even begin lactating, you should decide what you'll do with your milk and plan accordingly. Perhaps you'll keep some for family use and feed leftovers to pigs or poultry. Perhaps you'll begin experimenting with cheese, which sometimes uses a lot of milk for very little finished product. Perhaps you'll make goat's milk soaps and lotions (for a list of Web sites with recipes, see Appendix C). There are a myriad of options, but one you want to avoid is simply wasting the milk. In this time of peak oil and rising food costs, your milk can surely feed someone—even if it's your neighbor's livestock.

One thing to keep in mind: Regulations vary slightly between states, but if you are not a fully inspected, certified dairy, it's illegal for you to sell milk for human consumption. In some states, even if you're certified, selling raw (unpasteurized) milk is illegal, though in others it's fine as long as it doesn't cross state lines. When in doubt, check with your local extension agent or, if you're starting a commercial dairy, your state dairy inspector.

 Milk is filtered into quart jars in the kitchen of Ten Apple Farm in southern Maine. The stainless steel milk strainer is specifically designed to fit into wide-mouth quart or half-gallon jars.

That said, we all know farmers who leave a cooler and a jar on the front porch, and if that's within your comfort zone, it's certainly an option. At our house, we drink about two gallons of milk a week and make a lot of cheeses and yogurt for our own and our friends' consumption. I use fresh goat milk in cooking and baking—I love to surprise goat skeptics by revealing the secret ingredient! Karl takes a jar or two of milk to his office each week for colleagues to use in their coffee, and we happily give milk to friends; whatever surplus we're left with goes to the chickens. With just two lactating does, and a second refrigerator in the basement, we still become occasionally overwhelmed by the amount of milk in the house. Their wonderful efficiency makes goats prolific producers.

If you have just a goat or two and you're planning to keep and drink your goats' milk, before they kid you should invest in a stainless steel strainer, a box of filters, and at least a dozen half-gallon glass jars. We like to keep a dozen wide-mouth quart-size jars on hand too, so that we can store the milk on the lower shelves in the refrigerator. Strainers and filters are available from goat supply companies, and most will fit directly onto the opening of a wide-mouth jar.

If you have a larger herd, you'll want equipment that can keep up with a grander scale of milk production. You'll still need a strainer, but rather than straining directly into jars, you can use a seamless, stainless steel milk tote. If you're planning to pasteurize your milk, you may also want to invest in a small-scale home pasteurizer, available from most goat supply companies.

Before straining milk, you must sanitize or sterilize every piece of equipment that will touch it. This can either mean washing your strainer and jars in a bleach solution (a half ounce of bleach per gallon of water) or running them through the dishwasher. If you use bleach, make sure to rinse everything well, and either way, let the equipment air dry.

Straining the milk is relatively straightforward: Insert the filter into the strainer and secure it, rest the strainer on top of a sterilized jar or milk tote, and slowly pour the milk through the filter. Once it's strained, milk should be refrigerated immediately; the rule of thumb is to get milk to 38 degrees within an hour of milking. If you're storing milk in quart jars, a refrigerator will work, but if you're milking more than a few animals and storing the milk in larger quantities, it may be necessary to cool the milk in a water bath: Immerse the container of strained milk in a larger bucket of ice water. If you're getting more than five gallons of milk a day, consider buying a bulk tank or a water-immersion cooler, each of which will make the process of cooling more efficient.

If cooled properly, milk will keep in the refrigerator for five or six days, though the caproic acid in it gives a goatier flavor the older it gets. In our family, we usually hit our "goatiness threshold" for fluid milk at about three days, and use anything older for making cheese. We cover the tops of our jars with foil, secure it with the ring from a jar lid, and write the date and the milking time (a.m. or p.m.) on the foil so that there's never a mystery about its age. To make sure that milk isn't left to spoil in the back of the fridge, we also rotate the jars according to a "first in, first out" plan.

RAW OR PASTEURIZED?

Raw milk has received a lot of attention in recent years, and as you get started with dairy goats, you might wonder what the fuss is all about. Advocates for both raw and pasteurized milk can become impassioned in their arguments, and it can be hard to sort out the fact from the rhetoric. Simply put, raw milk is milk that is straight from the udder and has not been heat-treated. It may have been filtered to remove strands of hair or stray bits of dust, but that is the extent of its processing. Pasteurized milk, on the other hand, has been heat-treated. In the home dairy, this is usually done in a double boiler, where the milk is held at 145 degrees for thirty minutes, or 163 degrees for thirty seconds, and then immediately cooled to 38 degrees. Alternatively, milk can be heated in a home pasteurizer, which both heats and cools it.

Before the mid-1800s, all milk was consumed raw. Cheese and cultured products such as yogurt and kefir extended the life span of dairy products, as did rudimentary refrigeration, but in general, milk was drunk soon after it was expressed. In 1862, Louis Pasteur discovered that a relatively brief exposure to heat killed a majority of bacteria, molds, and yeasts. Once pasteurization was tried on milk, it was discovered that it radically extended the shelf life of dairy products, and gradually the practice became standard for commercially sold milk in Europe and the United States. Raw milk is still consumed in much of the world, though ultra-pasteurized milk that is shelf stabilized, such as Parmalat, is also widely available in developing countries.

Milk is sterile in the udder, but once it is expressed, and especially while it's still warm, it is an attractive incubator for a number of pathogens. These microorganisms will generally give milk an off flavor or will "clabber," or thicken and coagulate, it (the first step in ancient cheese making), but they can also remain in

 Bloomy rinded goat crottins age in a modern
cheese cave at Redwood Hill Dairy in
Sebastopol, California.

the milk without showing any signs. While straining milk removes visible bits of foreign matter, it does not guard against E. *coli* or *listeria* that can be present from manure dust. The pathogens that pasteurization removes also include the germs of brucellosis, diphtheria, scarlet fever, and tuberculosis, illnesses that were once common but, in developed countries, are now rare. For those who are pregnant, very young, or have a compromised immune system, even a slight exposure to any of these pathogens can represent a serious health risk.

The arguments in favor of raw milk are varied, but they include the milk being more natural; having a richer flavor, which is affected somewhat by pasteurization; and providing a homeopathic exposure to small amounts of the same pathogens that pasteurization eliminates. Raw milk advocates suggest that by eliminating even minimal exposure to bacteria and microflora, the human body loses its natural resistance to them.

In our family, we are not zealots, but we do drink raw milk from our goats and from the animals of farmers we know (whose sanitation methods we've seen). If I need heavy cream or cow's milk or even butter from the grocery store, I don't mind it being pasteurized. In fact, I prefer it, since I don't know the farmers but I do know the risks of raw milk. We feel comfortable drinking our goats' milk because we know that the animals are healthy, the conditions they live in are clean, the milking equipment is sanitized, and the milk is cooled quickly. There

 An old vanity plate on display at Redwood Hill Dairy in Sebastopol, California.

are farmers in our area whose raw milk I wouldn't drink, but there are plenty of farmers' whose raw milk I do drink.

When it comes to your own health and that of your family, the best plan is to make an informed decision and know that there's no right or wrong answer.

HEALTHY MILK

One of the best ways to tell if your lactating doe is healthy is to have a look at her milk. Healthy milk can sometimes take on the flavors of your doe's browse—beware of goats eating garlic scapes or onion greens—but it will always be bright white and fluid. Watch out for lumps or little chunks; a thick, almost ropey consistency; streaks of blood or pus; blood clots; or a bad smell (not the onions). If your doe's udder becomes hot to the touch or the tissue feels thickened, these are signs of mastitis, a severe infection that must be treated immediately by your veterinarian. Common treatment includes a round of antibiotics and irrigation of the mammary system, a process by which the udders are flushed with a penicillin solution. Milk from goats with mastitis is full of bacteria and should not be consumed by humans or fed to baby animals.

EASY CHEESE AND YOGURT

There is nothing more impressive to me than an interesting, well-crafted artisanal cheese. For thorough and well-written recipes for the home creamery, I would steer anyone to Ricki Carroll's excellent book, *Home Cheese Making*. For the novice, here are a couple of easy recipes that will still wow your friends:

Queso Blanco

This firm white cheese is popular in Latin America, and it makes an impressive appetizer if you're serving a Mexican meal. It doesn't melt, so it is also a great sub-

The finished product: Homemade bread and queso blanco cheese, mixed with dill in the foreground and plain in background.

stitute for tofu in stir-fries, and can be used in place of the Indian cheese *panir* in curries. It is somewhat bland and will absorb the flavors of the vegetables and spices it's cooked with. For an herbed cheese, mix a

Making cheese: After heating and curdling the milk, the whey is drained from the curd. The whey can be discarded, saved and made into ricotta, or fed to other livestock, such as chickens.

tablespoon or so of fresh or dried dill in the curd before hanging it. Because of the high cooking temperature, raw milk will become pasteurized during preparation.

1 gallon goat's milk

¹/₄ cup apple cider vinegar

In a large pot over medium-low heat, stir the milk until it reaches 180 degrees. Hold it at 180 degrees for 10 minutes, keeping a close eye on the temperature and continuing to stir to avoid scorching. Once milk is hot, its temperature climbs quickly, so you may need to lower the heat. Slowly stir in the vinegar and keep stirring until curd forms.

Line a colander with cheesecloth (a friend of ours uses clean, old T-shirts instead), set it in the sink, and slowly pour the curds and whey into the prepared colander. Tie the corners of the cloth into knots to form a bag, slip a wooden spoon handle under the knots, and hang the cheese inside the empty pot to drain. Let it hang for 5 to 6 hours, until it feels firm and stops dripping.

Remove from cloth and serve immediately, or refrigerate for later use.

Making cheese: The curd is hung to drain and dry, suspended in cheese cloth so that the whey drips into a stock pot.

Goat's Milk Yogurt

Goat's milk yogurt can become addictive, especially when it's sweetened with a little honey or blended with a mix of savory herbs. Our family eats yogurt in quantity, so we make large batches, often straining the evening's milk directly into the yogurt pot. Don't worry if it takes a few times to get it right—my first few experiments became "drinkable yogurt" or were used as salad dressing. Once you have a hand for it, yogurt is easy to make and will become a staple in your diet.

Yogurt is made from a live culture, and the easiest way to start is to buy a small cup of yogurt from the store. I often use cow's milk yogurt as a starter, and haven't noticed any great difference in flavor. (You can also buy yogurt starter culture from your local health food store.) To make the yogurt: Heat 1 quart of milk on the stove to 180 degrees, stirring regularly to prevent scorching. Remove the pot from heat and allow the milk to cool to 110 degrees. Stir about 1 cup of prepared yogurt into the milk, never allowing it to dip below 106 degrees, then either decant the mixture into clean cups in a commercially sold yogurt-maker, or cover and allow to ripen at 110 degrees for approximately 8 hours. (In the summer, we put ours in an oven that's been heated to 200 degrees and then immediately turn the oven off; in the winter, we put it behind the wood stove in the living room overnight.) Refrigerate the yogurt once it reaches a thick, custardy consistency. If the yogurt is too runny for your taste, you can thicken it by stirring in some powdered milk (goat's or cow's) before heating, approximately $^{1}/_{4}$ cup per quart.

CHAPTER NINE

Slaughter and Butchering

Beneath a tent of foil in a slow oven, a leg of kid and chunks of fennel bulb roast in a bath of white wine and Pernod, their savory anise perfume wafting through the house. On a weeknight, when time is short, goat rib chops braised with garlic and grape tomatoes make a quick meal. Finely chopped and blended with black truffles, artichoke hearts, liqueurs, and cream, an elegant goat meat pâté topped with onion marmalade will surprise and delight holiday guests. Gone are the days when goat meat was relegated exclusively to the domain of highly spiced curries and barbecues. While those are still great ways to enjoy the meat, goat has gone gourmet.

Never before has goat meat enjoyed a greater popularity in mainstream American cuisine. Available at butcher shops, some groceries, and farmers' markets throughout the country, goat meat has developed a following, and rightly so. Also called chevon, it's an exceptionally lean red meat, mild and meltingly tender when it's from a kid, stronger and a bit sinewy as the animal ages past a year. Goat is the most widely consumed red meat on the planet, and as a flavorful, environmentally light alternative to beef, its popularity will only grow.

To get that delectable roast, however, a goat must be slaughtered, and almost any farmer will tell you that deciding to end the life of one of

Dawud Uma, left, skins and guts one of the author's bucklings after slaughtering it according to *halal* ritual at a farm in southern Maine.

your animals is a challenge. Especially when it's a kid, so full of energy and life, it's a difficult choice, fraught with larger issues. What right do I have to decide this animal's fate? Will I be able to end a life that I helped bring into the world? How can I make the end of this creature's life most humane? These are just a few of the questions that most small farmers grapple with when deciding to cull an animal from the herd. If you're a carnivore, however, there's an undeniable honesty in making the choice to raise and slaughter your own animals. However uncomfortable the decision and the act may be, you know the source of your meat and you have an opportunity to be grateful to the animal.

At our farm, we dread slaughter day, but it also comes as something of a relief. Once we've made the decision to cull an animal, we find that the wait before slaughter is almost worse than the act. Whether you conduct the slaughter yourself, sell live goats to an individual buyer, or take them to auction, you'll notice that once they are actually gone from the farm, life resumes its normal rhythms. Making the choice isn't easy, but if you put some care into it you'll discover that, like so much on the farm, slaughtering and butchering one of your animals can be an enriching experience.

THE DECISION TO CULL

The first goats born on our farm were two bucklings. We had no desire to keep them as sires, and the idea of adding more wethers—mouths to feed that didn't earn their keep with any commodity—seemed foolish. We considered looking for a new home for them, and invited several animal lovers out to tour the farm and meet the kids. In the back of our heads, however, we kept ourselves open to the option of slaughtering them. We had heard horror stories from several farmers about selling bucklings as pets and having them returned when their new owners tired of them, sometimes after the goats had endured conditions that bordered on abuse. At the time, I couldn't imagine sending our bucklings into the unknown, but I also couldn't imagine eating these kids that I had helped deliver. (I have since reconsidered.) Knowing that our goats' meat would probably go to a Muslim family, we left the bucklings' horns and testicles intact so that the animals would be

zabiha, or unblemished, and appropriate for *halal* slaughter. Through a professor at the local university, we made contact with Dawud Uma, a former goat Before performing a *halal* slaughter, Dawud Uma ties three of the goat's legs together with baling twine. This practice helps keep the goat steady and prevents any damage to the animal and its carcass during the death throes.

farmer who was active in his mosque. To shield our two-year-old daughter from the process, we took the kids to a neighboring farm and met Dawud and another member of the Muslim community, who had come to assist. The girls and I said goodbye to the animals and left; Karl stayed with the two men and held the goats while their throats were slit with one quick stroke. The slaughter was performed on the final day of Ramadan, and once the meat had been butchered, it was distributed to needy families through an Islamic community center. Every part of the kids was used, from their meat to their hooves, horns, and hide. A Ghanaian drum maker is stretching the skins, and we hope someday to make music from our bucklings.

In the year of research that led up to our decision to raise goats, Karl and I had attended many slaughters and eaten more than our share of goat meat, but I use this story to illustrate the ambivalence that even the most committed goat meat enthusiast can feel at the idea of slaughtering one's own animals. It's a hard decision to make, and you should feel no embarrassment about acknowledging it as such. If you do choose to slaughter some of your animals, there are a few things you can do from the beginning to make the process a little less emotional:

Don't name the animal.

Though we know some farmers who name goats they know they'll slaughter, and actually label cuts of meat with their names, we find it much easier not to. While they're in our barn, we refer to them by physical characteristics (the bucklings were Blackie and Balls, for reasons that were obvious). We know other farmers who desensitize themselves a little by naming animals that are destined for the slaughterhouse after various foods: Chops, Suppertime, etc. Friends of ours who are raising a single steer named it Salisbury, as in steak.

If you have small children, don't let them get too attached.

At some point, human kids need to learn about mortality, but make sure that whatever story you stick with is age appropriate. Growing up on a farm is a great way for children to learn about the cycle of life, but too much, too soon can be traumatic. We don't believe in sugarcoating things, and our extra-carnivorous daughter, Charlotte, will figure things out soon enough. Until she begins to ask pointed questions, however, we tell her that the baby goats have moved to another farm.

Investigate your options.

Once you decide to cull a goat, there are several ways you can go about it. If you want to keep the meat for yourself, you can either slaughter the goat yourself or take it to be processed. If you want to sell the animal, you can arrange to have a buyer come to your farm and either slaughter it there or take it away live. You can also take it to auction, or team up with other local farmers and have a group of goats taken together. Whichever you decide on, make sure that you understand the process and are comfortable with all the steps and participants in it.

If you can, plan to be at the slaughter.

We have found that the anticipation of slaughtering our goats is much more tense and emotional than the act. This isn't because we're callous people; rather it's because in the moment, you're so focused on making the process go smoothly that an almost subconscious level of acceptance sets in. By providing the animal with a quick and respectful end of life, you are honoring its sacrifice.

Try to use all of its parts.

Whether you use cadaverous hooves to practice trimming, hollow the horn to make a trumpet, or tan the hide and hang it on your wall, there are virtually limitless uses for the inedible parts of the goat. Goats are so efficient in their lives that it's a shame to waste any part of them in death. A few phone calls before slaughter day should help you to find takers for the parts that you can't use.

DISPATCHING THE GOATS

You've made the difficult decision to cull a goat from your herd. The next step is choosing how the animal will be slaughtered. For the small-scale farmer, there are a few alternatives, each of which has its own advantages and drawbacks. If you don't want to keep the meat for your own consumption, you can sell the goat at auction, sell it live to an individual buyer, or have it slaughtered and butchered by a licensed, USDA-inspected facility, and then sell the meat. If you would like to keep the meat, you can take the animal, alive, to a butcher or packing house, hire someone to come to your farm to do it for you, or slaughter the animal yourself. If you're planning to sell the meat and there's a Muslim community in your area, you should look into your options for halal slaughter to ensure that the meat will be acceptable.

I'll talk about the pros and cons of each alternative, but keep in mind that in choosing your degree of removal from the process, the right decision is truly the one that feels most comfortable to you. While your animal's welfare includes a humane end of life, you should feel no obligation to put yourself in a situation that's too emotionally straining.

Auction

Once you've found an auction house that you like and figured out the logistics of transport, selling goats at auction is probably the easiest way to disperse of culled

HALAL SLAUGHTER
AND BUTCHERING

Recent immigrants from Africa and Asia make up a significant proportion of goat meat consumers in the United States. Among them, there are a large number of Muslims, and if you're planning to sell live goats or chevon to an Islamic community in your area, you should be familiar with the ideas of *zabiha* and *halal*, and how they relate in practical terms to your goats. Islamic law requires that an animal be zabiha, or unblemished, if its meat is to be eaten. This means that the horns and the testicles must be left on, and if it's a female, it cannot be pregnant. For a slaughter to be halal, or in compliance with Islamic law, it must be performed by a faithful Muslim in accordance with a specific set of guidelines that include facing Mecca, reciting a prayer over the animal, making sure that other animals are shielded from the sight of the slaughter, and slitting the animal's throat with one quick stroke. Halal slaughter is thoughtful and humane, and we know many non-Muslim farmers who take their animals to halal-certified slaughterhouses for ethical reasons.

We've found that reaching out directly to the local Muslim population through Islamic community centers and mosques is a wonderful way to connect with this market. And it's been a great way to make new friends!

 Goats are driven through livestock pens during the weekly goat and sheep auction at the Junction Auction House in Junction, Texas. While the majority of goat meat in the United States is still imported frozen from places like New Zealand, domestic production is on the rise, as evidenced by the growing number of goats going through places like Junction.

animals. Unless you live near one of the large goat and sheep auctions, like those in San Angelo, Texas, or New Holland, Pennsylvania, you'll probably deal with a small, family-run auction house. Some have sales specifically for goats and sheep; others have sales that include a little bit of everything—we've been to auctions where donkeys, geese, rabbits, and goats were for sale on the same morning. Goats sold at auction are generally sold either to meat packers or to livestock brokers who will resell them. Occasionally goats will go as pets, but usually if they sell at auction, they'll be slaughtered. The advantage to selling a goat at auction is that you're paid immediately and you don't need to coordinate anything beyond the animal's transportation to the auction. The disadvantages are that you don't know where the goat is going or the conditions in which it will be ending its life. Sale barns are notoriously rife with germs, and the goat that you've been keeping healthy this long may spend its last months in a much different state. You are also at the mercy of bidders, who buy the animals on the hoof and sometimes assess goats by lot rather than as individuals, so you can't set the animal's price.

Live sale at the farm

If there are members of your community with a taste for fresh goat meat, you may be able to arrange an individual sale. In this case, you'll sell the animal live and, generally, the purchaser will come to the farm and slaughter and butcher it there. One advantage to this option is that you can sell the goat for more per pound than you'll likely get at auction. Another advantage is that you can attend the slaughter, either as a participant or simply an observer. If a buyer is planning to slaughter a goat on your property, you'll need to prepare a space for the processing. The buyer should tell you what, specifically, they need, but it may include a rubber mat, plastic buckets to collect the blood and intestines, bags for the meat, and ice for chilling. If you're still working through your comfort with the slaughter, having it done on your property can also be the greatest disadvantage to a live sale.

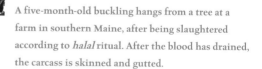 A five-month-old buckling hangs from a tree at a farm in southern Maine, after being slaughtered according to *halal* ritual. After the blood has drained, the carcass is skinned and gutted.

Selling meat butchered at a USDA-inspected slaughterhouse

To legally sell cuts of meat, you must have it slaughtered, butchered, and inspected by a USDA-certified packer. In rural areas and on the small scale, these are becoming few and far between, but they do exist. You may need to team up with other producers in your area to make the volume of goats worth a slaughterhouse's while, but if you choose this option, you can legally sell the meat in cuts at a farmers' market, or in quarters or halves to individual buyers. You can generally ask to be at the slaughter, though you won't be able to participate. One drawback to this option is that you must pay the facility and the inspector. If you're able to sell the meat for more by the pound at a farmers' market, it may justify the cost (usually calculated by a certain price per pound of the hanging weight). If the slaughterhouse in your area is not USDA certified, you can circumvent the law by selling the animal live to a customer, having them pay the butcher separately, and having them sign a waiver that specifies the meat is for their personal consumption (not sale).

For your own use

If you're keeping the meat for your own consumption, you can take the goat to a small slaughterhouse, where you'll pay a butchering fee and be asked to sign the same waiver mentioned above. Hiring someone to come to your farm to perform the slaughter will allow you to participate, but is not for the squeamish. Generally, the goat will be killed either with a gunshot (we've heard the expression "give him a shot of .22"), or by slitting the throat. Either way, the carotid arteries will be opened after death, and the animal will be suspended head-down to drain the blood before it is skinned, eviscerated, chilled, and finally butchered.

Your final option for home consumption is doing it yourself. I would not recommend this to the novice; there's simply too much room for error, especially if you don't have a more experienced hand to guide you. That said, if you have experience gutting and processing deer, the anatomy of a goat is similar enough that you can probably manage. Though I would still hesitate to recommend attempting a slaughter solo, Carla Emery's excellent *The Encyclopedia of Country Living*—a reference that should be on every small farmer's bookshelf—provides detailed, illustrated instructions.

CHEVON—THE OTHER RED MEAT

Once it's been butchered, you can anticipate that roughly one quarter of your goat's live weight will be meat. Basic butcher cuts for goat include ground, chops, ribs, roasts, steaks, stew meat, or bone-in leg and shoulder. Bones are often removed from the meat before it's packaged because they seem to intensify the strength of "goatiness." Goats have somewhat pointy bones, so be prepared for that if you do plan to cook the meat on the bone—the ribs, especially, can become little lances!

Ground or cubed as stew meat, chevon can be used interchangeably with beef or venison in many recipes, and until recently, many goat producers recommended using the meat predominantly for sausage making, or in chilis and barbecues. Lately, however, goat meat has gone upscale, and if you have an adventurous palate, I would recommend experimenting with recipes that explore the meat's natural flavors and affinities. Citrus, garlic, oregano, thyme, or virtually any pungent Mediterranean herb marries beautifully with the flavor of chevon, as do cinnamon, cumin, and dried fruits for a North African flavor. Unless it's from a very young kid, goat meat is at its best when cooked over low heat for a long time; a slow oven (250–300 degrees) or Crock-pot works well. To find more recipes, please consult the resource list in Appendix C.

Cuts of goat meat are tagged for distribution among several families after a *halal* slaughter in East Corinth, Maine.

CHEVON RECIPES

Our favorite recipes for chevon allow the meat's natural flavors to blossom by including spices that draw them out. The following roast leg of goat is an approximation of a meal we had at San Francisco's acclaimed One Market restaurant. It goes nicely with a mound of rice to absorb the juices, and a side of sautéed greens, either spinach or something sharper like mustard or broccoli rabe.

Roast Leg of Goat with Lemon and Herbs
Serves 4–6, depending upon the size of the leg

    1 leg of goat (3–4 pounds)
    6 cloves garlic (or more, to taste)
    $1/3$ cup olive oil
    coarse salt and freshly ground black pepper
    $1/4$ cup lemon juice
    1 lemon, sliced in rings
    large bunch oregano, on the stem
    large bunch thyme, on the stem
    water, white wine, or brown stock as needed

(continued)

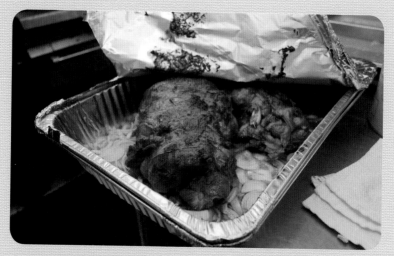

Roast goat, Hazon Food Conference, Isabella Freedman Retreat Center, Falls Village, Connecticut

1. Preheat oven to 300 degrees.

2. Pat dry the leg of goat. Slice 1 clove of garlic in half and rub it over the meat. Drizzle on 1 tablespoon of olive oil and season with salt and pepper to taste.

3. In a large casserole or roasting pan, brown the meat on both sides in 2 tablespoons of olive oil. Remove the pan from heat and pour the lemon juice and remaining olive oil over the meat. Tuck the lemon and herbs under the meat, and strew some over the top.

4. Cover and bake for 4 hours, checking periodically and adding water, white wine, or stock to make sure the pot stays moist. Roast is done when meat nearly falls from the bone and is tender when pierced with a fork.

Spiced Chevon Beggar's Purses

Our friend Leslie Oster, of Portland, Maine's Aurora Provisions, came up with these hors d'oeuvres. They're quick to make and a wonderful introduction to goat meat. Because chevon is so lean, they include a little extra rendered fat for sautéing the filling. This recipe makes 20 purses.

$^1/_4$ cup rendered duck or pork fat
3 cloves garlic, pressed
1 medium yellow onion, finely chopped
1 pound ground chevon
1 teaspoon cinnamon
$^1/_2$ teaspoon Chinese 5 spice powder
$^1/_2$ teaspoon turmeric
salt and freshly ground black pepper to taste
zest of one orange
1 package phyllo dough, thawed

1. Preheat oven to 350 degrees.

2. In a large frying pan over medium heat, melt the rendered fat. Add the garlic and onion and sauté until fragrant and translucent. Add the chevon, breaking it up with your spatula so that it browns evenly, and stir in the spices. Remove the pan from heat and stir in the orange zest.

3. Cut the sheets of phyllo dough in half, and place a moistened paper towel over them to keep them from drying out. Working quickly, loosen a sheet of dough from the stack and mound a heaping spoonful of meat into its center. Gather the edges of the dough together over the meat and pinch them so that they form a "purse."

4. Place the purses on a greased cookie sheet, leaving enough room between them so that they will become crispy and lightly browned. Bake for 15 minutes and serve hot.

CHAPTER TEN

Other Uses for Goats

Raising goats for milk and meat are just two of many paths you can take with these incredible animals. I touched on the subject briefly in the first chapter, but in this chapter I'll expand on the myriad options for those who are interested in small-scale goatkeeping. All the goat care that is discussed earlier in the book holds true for animals raised for any purpose, but for a more comprehensive discussion of fiber processing, goat packing, and construction of goat carts and chariots, please consult the resources listed in Appendix C.

MOHAIR AND CASHMERE

If you spin, knit, or felt, keeping goats to shear their mohair or comb out their cashmere is a wonderful way to follow the fiber from its rawest stage to its most finished. The depth of color in mohair that's dyed in small batches, the downy softness of minimally processed cashmere from a young kid: These are reason enough to raise goats for their coats. The satisfaction of wearing clothing made from the fleece of an animal that you've raised, or even delivered, is incredible.

One of Lani Malmberg's Ewe-4-ic Ecological Service weed-eating Cashmere goats eats the tall grass on the outskirts of Cheyenne, Wyoming. Lani says that she once tried to harvest the cashmere from her herd of over 1,200 goats, but it was so labor- and time-intensive that she hasn't tried again.

Santas with mohair beards (from Angora goats) in the gift shop at Austin's Mohair and Gifts in Harmony, Minnesota. Farmer and shop proprietor Ada Austin buys and sells crafts and clothing made by stay-at-home moms to help provide extra income for their families.

A detailed discussion of techniques for shearing mohair and combing cashmere is beyond the scope of this book, but if that's where your interest lies, there are a few goat management considerations that are specific to those who raise fiber animals. Goats with long, thick hair, such as Angoras, or with dense undercoats, like Cashmeres, can be raised in any climate, but will be most comfortable out of sweltering, humid heat while their coats are at their thickest. Most Angoras are shorn twice a year, before kidding in the spring and before breeding in the fall. Cashmere is generally combed in early spring, when the animal naturally starts to lose its winter coat and tufts begin to present themselves on the goat's outer "guard" hair. The coats of both types of goat are generally at their sparest during the summer, and in the United States the majority of Angoras and Cashmeres have traditionally been raised in the dry heat of South Texas. That said, if you're considering raising fiber goats in a part of the

Truffle, a naturally polled (hornless) Angora goat, stands in front of the barn at Friend's Folly Farm in Monmouth, Maine. In addition to raising and shearing Angora goats, owners Pogo and Marsha sell yarn and other fiber products and hold knitting workshops in a yurt on their farm.

country that gets brutal summer weather, be prepared to give your animals adequate shade and shelter, an abundance of water, and, if necessary, an early haircut.

External parasites, such as lice and mites, thrive on goats with thick fleece. There are numerous varieties of each, but most can be treated with a pour-on delouser, generally administered preventatively in two stages during the fall. If left untreated, goats can scratch themselves raw and end up with open sores and bacterial infections. The production of useable fiber can also drop by as much as a quarter. Check your goats periodically, and whether you use a commercial or a homemade remedy, be sure to treat them at the earliest sign of infestation.

If you plan to use or sell your goats' fleece, make sure to keep your animals clean throughout the year. It's much easier to pick out a few brambles here or there, or to cut out some matted spots every now and then, than to have to sort vigilantly through the fleece once it's been gathered. In commercial herds, fleece is graded and sorted, and any with urine stains, matting, or vegetation is discarded. A little attention throughout the year should keep as much of your goats' fiber as useable as possible.

Charlie Wilson and John Mionczynski lead a goat packing trip in the foothills of the Wind River Range near Lander, Wyoming. Mionczynski is considered by many to be the "father of American goat packing." Packing is a great way to enjoy goats and explore the outdoors in an easy and environmentally friendly way.

GOAT PACKING

My first true goat love was an Alpine pack goat from Wyoming named Panzer. Though he was not a master packer, he managed to carry his load and had a delightful, somewhat befuddled manner: He reminded me of an absent-minded professor, wandering off, investigating things, staring into space. On a five-day hike with him (and many other goats and people), I felt like we connected, and this attachment somehow left me feeling more attuned to the environment I was exploring.

Hitting the trail with a goat is an amazing experience, beyond anything I could have anticipated. A well-trained pack goat can carry up to a third of its body weight, will stay in close proximity to you without a leash, and will browse its way along the trail, requiring little (if any) supplemental food. Hiking with pack goats can feel almost decadent: You're along for a stroll, carrying as little as a daypack, while they do the heavy lifting, scrambling up the steepest path without complaint. As an environmentally light animal, goats leave little beyond their hoof prints and a scattering of their dung. As a companion, they stay near you, provide warmth on cold nights, and, if you're by yourself, their friendly and attentive nature keeps you from feeling lonely.

Karl and I have gone goat packing on organized trips twice, and liked it enough to start training two Alpine wethers of our own. The most thorough and extensive

 The goats of Wind River Pack Goats are transported in a trailer to the site of their next expedition, near Lander, Wyoming. Goats can be moved in any kind of livestock trailer, and sometimes, when necessary, in the back of an SUV or station wagon.

book about goat packing is written by the biologist and outdoorsman John Mionczynski and is called simply *The Pack Goat*. It is a tremendous resource, and the first purchase an aspiring goat packer should make—well before acquiring animals.

The most common goats to be used for packing are dairy wethers, though truly any goat can be trained. Dairy wethers are preferred because of their size— the larger the goat, the heavier pack it can carry. It's best to buy them when they're kids, and to bottle-feed them if possible so that they bond with you from an early age and consider you to be part of their herd. The horns of pack goats are generally left on so that they can efficiently cool themselves, though we know people who pack with disbudded goats, too. Saddles and panniers for packing are available from some retailers online, as are plans for making them yourself.

If you're interested in packing with goats, I would recommend going on an excursion with a company that leads trips so you can get a feel for the experience. If you decide to give it a try, before you buy your goats make sure that you have both time and land available for training.

Evin Evans of Split Creek Farm in Anderson, South Carolina, shows a Saanen doe at the 2004 American Dairy Goat Association National Show in Harrisburg, Pennsylvania. In addition to showing her own goats, Evans is an ADGA judge, traveling all over the country to assess all breeds of dairy goats in local shows.

SHOWING GOATS

Showing goats is a fun and rewarding pastime, and it's the gateway through which many small-scale farmers enter into the world of goats. Showing can instill confidence, especially in children, and the validation of seeing your goat place in competition is tremendous. Goat shows are typically held from late winter (in Florida) through the fall, often coinciding with fairs. In addition to the friendly competition of the ring, raising show goats connects you to other local breeders and gives you a community to turn to with questions and crises.

Goats can be shown through 4-H clubs and through various breed associations, with the exception of Kikos. Each association has its own breed standards and requirements for membership, and if you're interested in showing goats, you should look into the details of your preferred breed before acquiring animals. You should also be prepared for the expenses of breeding, grooming, and transporting goats to shows—depending upon the depth of your involvement, these can really mount.

GOAT CARTS AND CHARIOTS

My jaw dropped the first time I saw a grown man riding in a goat chariot, and I still get a kick out of goat races. The animals take off wildly, then come to an abrupt halt, sometimes inches before the finish line. Human participants in most races are forbidden to touch their animals on the track, so they plead and cajole and flap the reins uselessly until someone's goat crosses the line.

There is a long tradition of goat carts and chariots in the United States—photos exist of President Benjamin Harrison's children driving a goat cart across the White House lawn in the late 1800s—and it's continued to this day with organized (or, because of the goats, somewhat disorganized) races across the country. Many 4-H chapters build goat carts as group projects, and plans for construction are available through extension offices and on the Internet. As with goat packing, the most common animals used for racing are large wethers. Though it's not recommended to race with an animal that's younger than two years old, training with an empty cart should begin when they're kids so that the goats are comfortable with the equipment.

South African Boer goat judge Anton Ward questions 4-H kids about their animals during the showmanship competition of a sanctioned Boer goat show at the International Goat Days Family Festival in Millington, Tennessee. In showmanship, the animals aren't judged, but the kids are asked questions about their animals and required to demonstrate that they understand the proper way to handle and present them.

GOATS AS BRUSH CLEARERS

Goats are formidable weeders, and if you have an overgrown thicket or a brambly patch, they'll be delighted to clear it for you. In most cases, you need to do little but fence them into the right spot: Most noxious weeds are harmless to them and they'll generally avoid those that are poisonous. For reference, though, you should know that the following plants can be toxic to goats: black nightshade, bloodroot, bracken fern, delphinium, dock, hellebore, horse nettle, larkspur, lily-of-the-valley, locoweed, mayapple, milkweed, mountain laurel, nightshade, oleander, poison hemlock, pokeweed, rhubarb, sorrel, water hemlock, white clover, wilted leaves of any stone fruit such as peaches, plums, or wild cherries, and yew.

GOATS AS PETS

The emphasis in this book is on goats as livestock for the small-scale farmer, but on their own terms and with no expectations from them other than companionship, goats make excellent pets. While we are milking our Alpines and using them as a cornerstone of our modestly self-sufficient homestead, a family up the road from us enjoys the company of a single Nigerian Dwarf wether that lives in a pen next to their driveway. One or two goats can be wonderful companions for both humans and other animals: Racehorses are often stabled with a goat friend to keep them company. For human companions, pet goats are generally chosen from the smaller breeds, such as Pygmies, Nigerian Dwarfs, and Pygoras (a curly-locked cross of Pygmy and Angora), but this is purely for ease of containment. We've heard stories of goats being house-trained and even dressed up, so if the animal has the right temperament, you can have a lot of fun! Keep in mind that goats are herd animals, and they prefer to have constant company. Because of this, I would discourage intentionally getting a single goat. If a goat is alone, she'll adopt your family as her herd, but she'll be much happier with caprine companionship. If you do bring a single pet goat into your home, make sure that you can devote enough time to it that it's not lonely. Also keep in mind that though a goat's average lifespan is ten to twelve years, they can live as long as twenty.

Lee Waddell of Somerville, Tennessee, shakes the reins and yells at his goat, Rodeo, during the goat chariot races at the International Goat Days Family Festival in Millington Tennessee. The chariot races are the highlight of a goat-filled weekend that includes shows, milking demonstrations, and goat BBQ competitions.

AFTERWORD

Karl and I never thought we'd be farmers.
With liberal arts degrees and persistent cases
of wanderlust, we were urban and semi-nomadic before
the goats, settling in our adopted home of New York
City after covering, between the two of us, five conti-
nents. Now, much of our world revolves around ten rural acres. Our lifestyle of
subsistence farming and animal care couldn't be more different from our lives in
the city, and we rejoice in the change. At night, when the barn lights are off and
the kids, both human and hooved, are in bed, sometimes we go outside and stand
under the stars, inhaling the smell of our land, different in every season. We make
our livings at professions that are somewhat portable, but we are rooted in this little patch of earth, and it's goats that led us here.

That said, we still get the urge to travel, and we know a lot of goat owners with
farms of all sizes who manage the occasional getaway. Since goats live virtually
everywhere on the planet, you can include them in your plans, as we do, and visit
goat farms wherever you travel. The differences between a South African Boer
ranch and a hilltop Italian dairy are great, but from each we've brought back ideas for our own farm. And it feels good to have figured out a way to bring together the various, equally satisfying, pieces of our lives.

The author's Alpine dairy goat,
Flyrod, walking in shoulder
high snow at Ten Apple Farm,
in southern Maine.

 Charlotte leads a three-month-old
Alpine buckling kid on a leash at
Ten Apple Farm in southern
Maine.

Raising animals requires a large commitment
of all sorts of resources—emotional and economic—but it does not have to overwhelm you.
When you're deciding to raise goats, make sure
that you take your other interests into account and
set up your life in a way that doesn't sacrifice them.
What I mean by this is that you should decide at
the outset what scale of farm will allow you to get
enough rest, put aside money for other projects,

and keep things fun. This doesn't mean that every day will be easy, or that you'll never be surprised by an enormous vet bill or an expensive, unexpected barn repair. It just means that calamity should be the exception, not the norm. Before you bring any animals home, make certain that their upkeep will not be a financial strain. Your expenses will vary by scale, region, and management techniques, but once you've acquired your equipment, you should be able to calculate annual costs based on the prices of grain, hay, medication, breeding fees, and annual vet visits. Find a farm-sitter (if you don't know someone already, looking online is a great place to start) and add on a week of goat-sitting fees, just in case. The total cost of a year in goats may be larger than you imagined. While the animals are definitely worth the expense, make sure from the beginning that you don't stretch yourself too thin.

Even on a small scale, goats can offer tremendous learning opportunities for your family. Our animals help us impart daily lessons in responsibility, and as our girls get older, the goats will be instrumental in teaching them about the entire cycle of life, from reproduction to mortality.

If you bring goats into your life, they will change it profoundly. If you do it right—methodically, taking measured steps, refusing to drastically overextend your time and finances—caring for goats will change you for the better. They will delight you with their curiosity and their sense of adventure. They will test your mettle and teach you about your own capabilities. And, as you get to know them, they will display an astonishing amount of affection for you. It is hard not to anthropomorphize goats; their distinctive personalities and obvious intelligence lend them to description in human terms. If you're like me, you'll find yourself, at quiet times in the barn, talking to your goats like old friends.

APPENDIX A
The Year in Goats

JANUARY

The slowest month on the farm, January is the time to get organized. Compiling your records in a binder will help you keep track of breeding, vaccinations, feed, and expenses. It's also a good time to do research: If there's a Muslim community in your area, find out when Ramadan falls this year. Bone up on your wild plant identification. Talk to your extension agent about any questions you have for the coming year. Periodically check your goats' shelter to make sure that it's holding up against winter weather. If you're ordering seeds for spring planting, think about growing a goat garden.

FEBRUARY

If you bred your does in September, kidding season will start this month. Make sure that your kidding kit is well stocked and easily accessible at least two weeks before the first kid is due. Keep a heat lamp on hand. Also, inform your vet or goat mentor, and make sure you have their after-hours phone number. (Keep in mind that your busiest time of year is also your mentor's, so try to be considerate of his or her time.) If you plan to milk your goats and are new to it, order any supplies you may need so that you have them at the ready. In most parts of the country, this will be a month when there's little forage available outside, so check your hay supply to make sure it will last—if in any doubt, call your hay supplier as soon as possible. Start checking your fencing to make sure it's sound so that you can put the goats out whenever weather allows.

An angora goat grazes outside a window at MyTime Ranch, Eureka, California.

MARCH

Depending upon where you live, your Cashmere goats may begin to shed their fleece. This could happen as early as the end of this month, or as late as May. Be sure to have your equipment ready for gathering. If you already have kids, monitor their growth to make sure they're in good shape. If you bred in October, keep that kidding kit handy. It's time to think about deworming for the whole herd, but especially for any new mothers. Make sure that your pens and pastures aren't too damp and muddy—if there's standing water, see about getting some clean fill. Also make sure that your goats' hooves are nicely trimmed to avoid any chance of hoof rot. If there's an Easter market for goat meat in your area, make contact with the community to let them know that you have kids for sale.

APRIL

It's getting warmer and the goats are spending more time outdoors. While they nibble at the tender new plants, watch their bellies for bloat and their poop for any sign of scours. If you've been putting off that thorough spring cleaning of the barn, April is the time to do it—waiting much longer will give you stinky pens and grouchy goats. Dedicate a compost pile to goat manure, for later use in the garden. Angora goats can be shorn now, so make sure that you have equipment for harvesting the fleece.

MAY

The goats are outside almost all the time now. The last of the kids should be born, and if you're milking, you'll have an abundance—it's a good time to experiment with cheese. Farmers' markets are starting up for the season, and it's time to get in touch if you're interested in participating. If you show goats, make sure your animals are in compliance with all the rules for eligibility before the season begins. Workshops to train goat judges are often held in May—it might be fun to attend or audit.

JUNE

It's summer now, so make sure that your goats have plenty of fresh, clean water. Clean their buckets daily to prevent the buildup of slime. If you have the space to rotate their pasture, do it regularly to keep them eating a variety of browse. Make sure that their shelter provides enough shade. It's a good time of year to go goat packing, so gather your goats and hit the trail! Also call your hay supplier to see what's available. If you plan to buy your hay in the field, this may be the month. If you haven't already, plant your goat garden this month so it's ready in the fall.

July

What better way to celebrate Independence Day than with a goat race? Harness the chariot, mark a course, and invite your neighbors for a day at the races—with a little chevon barbecue and goat's milk ice cream, of course. Keep an eye on shelter and shade, and make sure that the goats are hydrated.

August

Show season is in full swing, and if you show your goats, you may feel like you're never home this month. At the end of summer, you should look at your herd and evaluate the goats, deciding how many you'll keep through the winter. Find an outlet for the animals if you're going to cull, either through sale or slaughter. Start calculating your hay needs and make sure that your supplier knows how much you'll be taking.

September

September is a busy month on the farm. If you show your goats, there are still fairs through the month. The International Goat Days Family Festival in Millington, Tennessee, is a great showcase for all things goat—especially goat chariot races! In many parts of the country, the beginning of cold weather will bring does into heat, so it may be time to breed. Start flushing does three weeks before you plan to breed them. Make sure that you've got a buck (or an AI appointment) lined up for each doe as soon as she's ready, and record the dates of conception and estimated birth. If you keep a buck, make sure your fences are in great shape to avoid escapes and accidental breeding. Now is the time to think about autumn deworming. Your goats will still be browsing outside most of the time—keep them away from the wilted leaves of any stone fruit trees, like peach, plum, or cherry.

October

Breeding season is in full swing and forage is becoming sparse. As you're pulling out your garden, feed the goats stalks and stems, but be sure they don't overdo it on the fruit and brassicas (plants in the cabbage family)—either can make a goat sick. Shear mohair before the weather gets too cold so that the goats don't get chilled. Your winter hay should be in storage by the end of the month, and now's the time to do a careful barn cleaning, bringing in any medicines and equipment that should be kept from freezing. If you can get away, the American Dairy Goat Association's annual convention is at the end of the month.

November

This is the month to get buttoned up for winter. Check your barns and shelters to make sure they're not drafty, and put down a little extra fresh bedding to insulate for the season. Keep water buckets from freezing, either by filling them regularly with warm water or by investing in an electric water heater. As long as the goats can go outside comfortably, keep them exercising, especially the pregnant does. If there are hunters near your pastures, check that your property lines are clearly marked and keep an eye on the goats to make sure they don't get spooked.

December

If there are any does you haven't bred, this month will probably be their last cycle. Other than the year's final breeding, December is a month for maintenance. Keep an eye on your hay stores and make sure that you haven't underestimated your needs. Check the pens to keep them snug and comfortable. If you have lactating does due in February, gradually dry them off so that they have two good months to rest. If there's snow, why not strap on your snowshoes and take the pack goats for a walk? And a chilly winter afternoon is the perfect time to slow-roast a leg of goat!

APPENDIX B

Equipment and Essentials by Chapter

Chapter One: Things to Think About in Advance
For an overview of virtually every breed of goat known to man, visit Oklahoma State University's excellent compendium of livestock breeds at ansi.okstate.edu/breeds/goats/.

Chapter Two: Where Your Goats Will Live

Shelter
Be it a large barn or a three-sided shed, your goat shelter needs to be sturdy and appropriate to your climate, preferably with at least two enclosed (and lockable) pens.

Pitchfork, Shovel, and Wheelbarrow or Oxcart
These don't have to be fancy, but you will need tools to pry apart layers of bedding, remove it from the pens, and transport it to your manure pile.

Bobcat or Tractor
These should not replace the tools for manual pen cleaning, but they'll make the job much easier.

Shavings, Sawdust, or Straw
Shavings and sawdust are absorbent and easy to clean up from the pen, while straw keeps the bedding from sticking to the goats. We get large bags of sawdust from a woodshop near Karl's office to supplement the shavings and straw that we buy from our local farm store. All three expand when they're released from their bags or bales, so don't open every package at once—you may end up with more than you need.

Agricultural Lime
This is sold in twenty-five and fifty-pound bags, and you'll sprinkle it liberally on the barn floor to keep the flies down and the pens smelling sweet. Make sure that you ask for agricultural lime at the farm store, rather than hydrated lime, which is caustic.

Broom
A little cleanup goes a long way. Dedicate a broom to the barn and sweep regularly to keep stray shavings and hay from building up. Not only does the barn look more orderly, but a clean barn is less of a fire hazard.

CHAPTER THREE: THE ETERNAL PROBLEM OF GOAT FENCING

Everyone's fencing requirements are different, and you should look at your land and your animals to evaluate your needs. Your local farm store is a great place to start, but many don't stock a wide variety of electrified or woven wire fencing. The catalog and Web site for Premier 1 are wonderful sources for fencing, and they can help you decide the right system for your property and budget. Your extension agent can also be a resource, though ours actually sent us to Premier! The options are many, but make sure that your fence is strong and high enough to deter the goats.

ENERGIZER

If you buy electrified fencing, make sure that you don't forget to order the energizer (battery), which is generally sold separately. If your region gets a lot of sun, consider a solar option.

FENCE TESTER

For $15, it takes the mystery out of the electric fence.

CHAPTER FOUR: FEEDING YOUR GOATS

HAY

Good quality, second-cut hay is best. Keep bales off the ground and out of direct sunlight. Cover stored bales with a tarp so they don't get dusty.

HAY FEEDER

Goats are notorious hay wasters, so get a feeder that gives them a little resistance and place a tray or a lipped shelf underneath it to catch stray bits.

GRAIN

Premixed grain with appropriate minerals for your area is available from most farm stores in fifty-pound bags. Make sure it's fresh (no dusty bags) and not medicated.

GRAIN PANS

Heavy duty rubber pans work well. If you only have a few goats, you can buy small pans and feed them each individually, which cuts down on the suppertime jostling.

WATER BUCKETS

Five-gallon plastic buckets are available at farm and hardware stores. Keep a few spares around—they're good for all kinds of things. Leave them clean, thawed (in winter), and always full.

MINERAL BLOCK

Mineral blocks for horses are just fine for goats, and can be either placed on the ground in the paddock or attached to the wall of the pen. If you house goats and sheep together, keep the mineral block out of sheep's reach because it contains too much copper for their systems.

Scoop

Old-timers use coffee cans; we prefer empty two-pound yogurt containers.

Chapter Five: On Bucks and Breeding

Calendar

Dedicate a separate calendar to heat cycles, breeding dates, and projected due dates—otherwise, goat reproduction is likely to take over the family calendar.

Chapter Six: Raising Kids

Kidding Pen

Make sure you have an empty pen for the laboring doe and her new kid. You may move her there immediately before birth, or decide to isolate her earlier.

Extra Hay Feeders, Grain Pans, and Water Buckets for Kidding Pen

Get the pen ready beforehand so you know you aren't missing something.

Kidding Kit

Buy pre-assembled or put together your own, but make sure it's ready at least two weeks before the first kid is due.

Molasses

Warm molasses water helps a doe replenish her system after giving birth. Depending on the size of your herd, you can buy animal grade molasses at your local farm store or a big jug from the grocery.

Appliance Box

We've all done it. At some point a weak kid will end up in your kitchen, so before kidding season, find a big cardboard box, just in case. Appliance stores are a great place to look— you can either ask them to put one aside, or look out behind the store.

Powdered Colostrum and Milk Replacer

It's good to keep these on hand, just in case. Make sure they are milk based rather than soy.

Empty Soda Bottles and Rubber Nipples

These should be part of your kidding kit and should be clean and ready for use.

Tattoo Kit

This should include tattoo tongs, ink, numbers, and letters. Make sure to sterilize everything before and between uses, and to test each tattoo on a piece of paper before applying it to your goat.

Disbudding Iron
If you're planning to disbud the kids, make sure you have access to a disbudding iron in their first ten days. These are available from goat supply companies.

Emasculator or Elastrator and Bands
For castrating bucklings, these are available from most livestock supply companies.

Chapter Seven: The Health of Your Goats

Hoof Trimmer and Plane or Rasp
These can be purchased from goat supply companies, or simply use sharp pruning shears with short, straight blades. Hoof trimming kits can also be purchased, and often include a hoof plane, hoof pick, and antibacterial ointment.

Dewormer
Once you and your vet have determined an appropriate deworming plan, you can buy most dewormers through livestock supply companies. There are many alternatives, so make sure that you and your vet choose one that targets the parasites specific to your environment.

CD/T
When you're comfortable giving shots, this essential vaccine should be kept on hand for new kids and your goats' annual booster. Available from livestock supply companies, it must be refrigerated, but will keep for a couple of years.

Emergency Kit
Stock your kit with more than you think you might need, including syringes and needles, scalpel, antibacterial spray, surgical scissors, gauze, tape, and a flashlight. These can be purchased ready-assembled from goat supply companies, or you can put one together yourself.

Chapter Eight: Goat Milk?

Clean Rags
We like old dishtowels, but anything soft will do.

Antibacterial Udder Solution or Udder Wipes
These are available commercially prepared from goat supply companies. Recipes for home-made solutions can be found at fiascofarm.com/dairy.

Teat Dip and Cup or Film Canister
Cow teat dip is fine and available at most farm stores.

Udder Cream
Products for cows are fine—we like Udderly Smooth, but there are many others on the market. Keep this handy, especially right after kidding, when does are likely to get chapped teats.

Strip Cup

A small tin or plastic cup with a handle and a screened lid, a strip cup catches the first few squirts of milk and helps you check for early signs of mastitis. They are available from dairy supply companies, or can be homemade with a bit of tightly woven screen and a cup.

Milking Machine and Vacuum Pump

Available new through goat supply companies, but often can be found used. If you're new to goats, make sure that you read the directions carefully, sterilize all equipment, and, if possible, have a more seasoned farmer show you how it works!

Seamless, Stainless Steel Milk Pail

These usually cost about $50 and are available from your farm store or any dairy supply company. They should be sanitized between milkings.

Hanging Scale

Great for weighing milk and, if you get a special sling, for weighing new kids. Available at most farm stores; hang the scale from a rafter near the milk stand for easy access.

Milk Tote

Not essential, but very handy if you have a long walk back from the barn. Seamless, stainless steel totes are best, and are available from dairy supply companies. You can strain milk directly from the pail into the sterilized tote and put a lid on before leaving the barn, so you don't have to worry about sloshing.

Strainer and Filters

Essential if you're planning to drink your milk, a strainer and disposable dairy filters catch any bits of hair or shavings. Most fit directly onto a milk tote or wide-mouth jar, and are small enough to sterilize in the dishwasher.

Home Pasteurizer

Shaped like a Crock-pot, a home pasteurizer heat-treats roughly two gallons of milk at a time. Some shut off automatically when milk is pasteurized; others also contain a cooling system.

Quart and Half-gallon Glass Jars

If you only have a few goats, keep sterilized jars on hand for storing milk.

CHAPTER NINE: SLAUGHTER AND BUTCHERING

Freezer

Equipment for the actual slaughter will vary depending upon how and where it's done. If it will be done on your property, discuss your butcher's needs well beforehand. If you are keeping the meat, it's essential that you have adequate freezer space. Calculate amount of meat to be one quarter of goat's live weight.

APPENDIX C
Resources and Suppliers

BOOKS

The following is a list of books that we've found helpful around the homestead. Some deal exclusively with goats and some offer more general information, but we've found all to be valuable.

Amundson, Carol. *How to Raise Goats*. Minneapolis: Voyageur Press, 2008.

Belanger, Jerry. *Storey's Guide to Raising Dairy Goats*. North Adams, MA: Storey Publishing, 2001.

Caldwell, Gianaclis, and Ricki Carroll. *Mastering Artisan Cheesemaking: The Ultimate Guide for Home-Scale and Market Producers*. White River Junction, VT: Chelsea Green Publishing, 2012.

Carroll, Ricki. *Home Cheese Making*. North Adams, MA: Storey Publishing, 2002.

Ciletti, Barbara. *Making Great Cheese at Home*. New York: Lark Books, 1999.

Crepin, Joseph. *La Chevre*. Philo, CA: Mountain House Press, 1990.

Damerow, Gail, ed. *Barnyard in Your Backyard*. North Adams, MA: Storey Publishing, 2002.

Eberhardt, J. E. *Good Beginnings with Dairy Goats*. Medford, WI: Dairy Goat Journal, 1985.

Ekarius, Carol. *How to Build Animal Housing*. North Adams, MA: Storey Publishing, 2004.

—*Small-Scale Livestock Farming: A Grass-Based Approach for Health, Sustainability, and Profit*. North Adams, MA: Storey Publishing, 1999.

Emery, Carla. *The Encyclopedia of Country Living, Tenth Edition*. Seattle: Sasquatch Books, 2008.

Kinsedt, Paul, with the Vermont Cheese Council. *American Farmstead Cheese: The Complete Guide to Making and Selling Artisan Cheeses*. White River Junction, VT: Chelsea Green Publishing, 2005.

Luttman, Gail. *Raising Milk Goats Successfully*. Charlotte, VT: Williamson Publishing, 1986.

Mionczynski, John. *The Pack Goat*. Boulder, CO: Pruett Publishing Company, 1992.

Pugh, D. G. *Sheep and Goat Medicine*. St. Louis: Saunders, 2001.

Roberts, Jeffrey. *Atlas of American Artisan Cheese*. White River Junction, VT: Chelsea Green Publishing, 2007.

Sayer, Maggie. *Storey's Guide to Raising Meat Goats*. North Adams, MA: Storey Publishing, 2007.

Weaver, Sue. *The Backyard Goat: An Introductory Guide to Keeping Productive Pet Goats*. North Adams, MA: Storey Publishing, 2011.

Weaver, Sue. *Goats: Small-Scale Herding for Pleasure and Profit*. Irvine, CA: Hobby Farm Press, 2006.

ORGANIZATIONS

Most goat organizations are run by volunteers and exist without permanent office space. Unless otherwise indicated, it's easiest to contact each association through its Web site.

DAIRY

American Cheese Society: cheesesociety.org

American Dairy Goat Association: adga.org

FIBER

American Angora Goat Breeders Association: aagba.org

Cashmere Producers of America: capcas.com

Colored Angora Goat Breeders Association: cagba.org

Eastern Cashmere Association: easterncashmereassociation.org

Northwest Cashmere Goat Association: nwcashmere.org

MEAT

American Boer Goat Association: abga.org

American Kiko Goat Association: kikogoats.com

International Boer Goat Association: intlboergoat.org

International Kiko Goat Association: theikga.org

United States Boer Goat Association: usbga.org

ALL BREEDS

American Livestock Breeds Conservancy: albc-usa.org

The International Goat Association: iga-goatworld.org

GENERAL INFORMATION

americangoatfederation.org

fiascofarm.com

fiberart.com

kysheepandgoat.org

sheepandgoat.com

SUPPLIERS
Billy Goat Gruff (gates and portable pens)
(800) 733-4283
tartergate.com
Caprine Supply
(913) 585-1191
caprinesupply.com

Dairy Connection
(608) 242-9030
dairyconnection.com

Glengarry Supply
(888) 816-0903
glengarrycheesemaking.on.ca

Hamby Dairy Supply
(800) 306-8937
hambydairysupply.com

Hoegger Supply Company
(770) 461-5398
thegoatstore.com

Jeffers Livestock Supply
(800) 533-3377
jefferslivestock.com

New England Cheesemaking Supply Company
(413) 628-3808
cheesemaking.com

Premier 1 (fencing)
(800) 282-6631
premier1supplies.com

APPENDIX D

State Cooperative Extension Offices

Also see the United States Department of Agriculture Web site at csrees.usda.gov/Extension/index.html.

ALABAMA
Alabama Cooperative Extension System
109-D Duncan Hall
Auburn University, AL 36849
(334) 844-4444
aces.edu

ALASKA
University of Alaska–Fairbanks
Cooperative Extension Service
308 Tanana Loop, Room 101
P. O. Box 756180
Fairbanks, AK 99775-6180
(907) 474-5211
alaska.edu/uaf/ces

ARIZONA
University of Arizona
Cooperative Extension
Forbes 301
P.O. Box 210036
Tucson, AZ 85721-0036
(520) 621-7205
ag.arizona.edu/extension

ARKANSAS
University of Arkansas
Division of Agriculture
Cooperative Extension Service

2301 South University Avenue
Little Rock, AR 72204
(501) 671-2000
uaex.edu

CALIFORNIA
University of California, Davis
Agriculture and Natural Resources
Cooperative Extension
One Shields Avenue
Davis, CA 95616-8521
(530) 752-1250
http://ucanr.org/ce.cfm

COLORADO
Colorado State University Extension
Campus Delivery 4040
Fort Collins, CO 80523-4040
(970) 491-6281
ext.colostate.edu/

CONNECTICUT
University of Connecticut
Cooperative Extension System
W. B. Young Building, Room 231
1376 Storrs Road, Unit 4134
Storrs, CT 06269-4134
(860) 486-9228
extension.uconn.edu

DELAWARE
University of Delaware
College of Agriculture and Natural
Resources
Cooperative Extension, Townsend Hall
531 South College Avenue
Newark, DE 19716-2103
(800) 282-8685
http://ag.udel.edu/extension/index.php

FLORIDA
University of Florida
IFAS Cooperative Extension
P. O. Box 110210
Gainesville, FL 32611
(352) 392-1761
http://solutionsforyourlife.ufl.edu

GEORGIA
University of Georgia
College of Agriculture and Environmental
Sciences
Cooperative Extension
111 Conner Hall
Athens, GA 30602
(800) ASK-UGA1
ugaextension.com

HAWAII
University of Hawaii
College of Tropical Agriculture and Human
Resources
Cooperative Extension Service
3050 Maile Way, Gilmore 203
Honolulu, HI 96822
(808) 956-8139
ctahr.hawaii.edu/site/Extprograms
.aspx

IDAHO
Director of Cooperative Extension
University of Idaho at Twin Falls
P. O. Box 1827
Twin Falls, ID 83303-1827
(208) 736-3603
extension.uidaho.edu

ILLINOIS
University of Illinois Extension
Office of Extension and Outreach
214 Mumford Hall MC-710
1301 W. Gregory Drive
Urbana, IL 61801
(217) 333-5900
http://web.extension.uiuc.edu/state/index
.html

INDIANA
Purdue University Extension
Animal Science Department
915 W. State Street
West Lafayette, IN 47907-2054
(765) 494-4808
ces.purdue.edu/counties.htm

IOWA
Iowa State University
University Extension
2150 Beardshear Hall
Ames, IA 50011
(515) 294-6675
extension.iastate.edu

KANSAS
Kansas State University
115 Waters Hall
Manhattan, KS 66502
(785) 532-6147
oznet.ksu.edu

KENTUCKY
University of Kentucky
College of Agriculture
S-107 Ag. Science Building, North
Lexington, KY 40546
(859) 257-4302
http://ces.ca.uky.edu/ces

LOUISIANA
Louisiana State University
Agricultural Center
101 Efferson Hall
P. O. Box 25203
Baton Rouge, LA 70803
(225) 578-4161
lsuagcenter.com/en/our_offices/
parishes

MAINE
University of Maine Cooperative Extension
5741 Libby Hall
Orono, ME 04469-5741
(207) 581-3188
umext.maine.edu

MARYLAND
University of Maryland
College of Agriculture and Natural Resources
1413A-AnSc/AgEn Building
College Park, MD 20742
(301) 405-1366
http://extension.umd.edu

MASSACHUSETTS
Agriculture and Landscape Extension
French Hall
230 Stockbridge Road
University of Massachusetts
Amherst, MA 01003-9316
(413) 545-0895
umassextension.org

MICHIGAN
Michigan State University Extension
Agriculture Hall, Room 108
Michigan State University
East Lansing, MI 48824-1039
(517) 355-2308
msue.msu.edu

MINNESOTA
University of Minnesota Extension Service
240 Coffey Hall
1420 Eckles Avenue
St. Paul, MN 55108-6068
(612) 624-1222
extension.umn.edu

MISSISSIPPI
Mississippi State University
P. O. Box 9601
Mississippi State, MS 39762
(601) 325-3036
http://msucares.com

MISSOURI
University of Missouri
College of Agriculture
2-28 Agriculture Building
Columbia, MO 65211
(573) 882-6385
http://extension.missouri.edu

MONTANA
MSU Extension
Montana State University
203 Culbertson
P. O. Box 172230
Bozeman, MT 59717-2230
(406) 994-1750
http://extn.msu.montana.edu

NEBRASKA
University of Nebraska Extension
211 Agriculture Hall
Lincoln, NE 68583
(402) 472-2966
extension.unl.edu

NEVADA
University of Nevada, Reno
Cooperative Extension
Mail Stop 404
Reno, NV 89557
(775) 784-7070
unce.unr.edu

NEW HAMPSHIRE
UNH Cooperative Extension
200 Bedford Street (Mill #3)
Manchester, NH 03101
(877) 398-4769
http://extension.unh.edu

NEW JERSEY
Rutgers, The State University of New Jersey
Cooperative Extension Service
88 Lipman Drive
New Brunswick, NJ 08901
(732) 932-7000 ext. 4204
http://njaes.rutgers.edu

NEW MEXICO
New Mexico State University
Extension Animal Sciences
Knox Hall, Room 232
P. O. Box 30003 MSC 3AE
Las Cruces, NM 88003
(505) 646-3326
cahe.nmsu.edu

NEW YORK
Cornell University
Cooperative Extension
Box 26—Kennedy Hall
Ithaca, NY 14853
(607) 255-0789
cce.cornell.edu

NORTH CAROLINA
North Carolina A & T State University
Coltrane Hall
Greensboro, NC 27411
(336) 334-7956
ag.ncat.edu/extension

NORTH DAKOTA
North Dakota State University
Extension Service
NDSU Dept. 7000
315 Morrill Hall
P.O. Box 6050
Fargo, ND 58105-6050
(701) 231-8944
ext.nodak.edu

OHIO
Ohio State University
Cooperative Extension
2120 Fyffe Road
Room 3 Ag Admin
Columbus, OH 43210
(612) 292-6181
http://extension.osu.edu

OKLAHOMA
Oklahoma State University
Cooperative Extension Service
Division of Agricultural Sciences
136 Ag Hall

Stillwater, OK 74078
(405) 744-5398
oces.okstate.edu

OREGON
Oregon State University
Extension Service Administration
101 Ballard Extension Hall
Corvallis, OR 97331-3606
(541) 737-2713
http://extension.oregonstate.edu

PENNSYLVANIA
Pennsylvania State University
Cooperative Extension Service
114 Ferguson Building
University Park, PA 16802
(814) 865-1688
extension.psu.edu

RHODE ISLAND
University of Rhode Island
Cooperative Extension
Fisheries/Animal/Vet Science Department
210B Woodward Hall
Kingston, RI 02881
(401) 874-2900
uri.edu/ce/

SOUTH CAROLINA
Clemson University
Cooperative Extension
103 Barre Hall
Clemson, SC 29634-0110
(864) 656-3382
http://clemson.edu/extension

SOUTH DAKOTA
South Dakota State University
Cooperative Extension Service

AGH 154 /Box 2207D
Brookings, SD 57007
(605) 688-4792
http://sdces.sdstate.edu

TENNESSEE
University of Tennessee
Extension Service
2621 Morgan Circle
121 Morgan Hall
Knoxville, TN 37996
(865) 974-7115
utextension.utk.edu

TEXAS
Texas A & M University
Cooperative Extension Service
106 Jack K. Williams Administration Bldg.
7101 TAMU
College Station, TX 77843
(979) 845-7800
http://texasextension.tamu.edu

UTAH
Utah State University
Extension Service
Animal, Dairy, and Veterinary Sciences
4815 Main Hill
Logan, UT 84322
(435) 797-2162
http://extension.usu.edu

VERMONT
University of Vermont
Extension Service
19 Roosevelt Highway, Suite 305
Colchester, VT 05446
(802) 656-2990
uvm.edu/~uvmext

VIRGINIA
Virginia Tech
Cooperative Extension Service
101 Hutcheson Hall (Mail Code 0402)
Blacksburg, VA 24061
(540) 231-5299
ext.vt.edu

WASHINGTON
Washington State University
Extension Service
P. O. Box 646248
Hulbert 411
Pullman, WA 99164-6248
(509) 335-2837
http://ext.wsu.edu

WEST VIRGINIA
West Virginia University
Extension Service
Room 817 Knapp Hall
P. O. Box 6031
Morgantown, WV 26506-6031
(304) 293-5691
wvu.edu/~exten

WISCONSIN
University of Wisconsin Extension
Agriculture and Natural Resources
633 Extension Bldg
432 N. Lake Street
Madison, WI 53706-1498
(608) 263-7320
uwex.edu/ces

WYOMING
University of Wyoming
Cooperative Extension Service
Department 3354
1000 E. University Avenue
Laramie, WY 82071
(307) 766-5124
http://ces.uwyo.edu